Lecture Notes in Computer Science 15013

Founding Editors

Gerhard Goos
Juris Hartmanis

Editorial Board Members

Elisa Bertino, *Purdue University, West Lafayette, IN, USA*
Wen Gao, *Peking University, Beijing, China*
Bernhard Steffen, *TU Dortmund University, Dortmund, Germany*
Moti Yung, *Columbia University, New York, NY, USA*

The series Lecture Notes in Computer Science (LNCS), including its subseries Lecture Notes in Artificial Intelligence (LNAI) and Lecture Notes in Bioinformatics (LNBI), has established itself as a medium for the publication of new developments in computer science and information technology research, teaching, and education.

LNCS enjoys close cooperation with the computer science R & D community, the series counts many renowned academics among its volume editors and paper authors, and collaborates with prestigious societies. Its mission is to serve this international community by providing an invaluable service, mainly focused on the publication of conference and workshop proceedings and postproceedings. LNCS commenced publication in 1973.

Marco Piangerelli · Bardh Prenkaj ·
Ylenia Rotalinti · Ananya Joshi · Giovanni Stilo
Editors

Discovering Drift Phenomena in Evolving Landscapes

First International Workshop, DELTA 2024
Barcelona, Spain, August 26, 2024
Proceedings

Editors
Marco Piangerelli ⓘ
University of Camerino
Camerino, Italy

Bardh Prenkaj
Technical University of Munich
Munich, Germany

Ylenia Rotalinti
Brunel University
London, UK

Ananya Joshi
Carnegie Mellon University
Pittsburgh, PA, USA

Giovanni Stilo ⓘ
University of L'Aquila
L'Aquila, Italy

ISSN 0302-9743   ISSN 1611-3349 (electronic)
Lecture Notes in Computer Science
ISBN 978-3-031-82345-9   ISBN 978-3-031-82346-6 (eBook)
https://doi.org/10.1007/978-3-031-82346-6

© The Editor(s) (if applicable) and The Author(s), under exclusive license to Springer Nature Switzerland AG 2025

This work is subject to copyright. All rights are solely and exclusively licensed by the Publisher, whether the whole or part of the material is concerned, specifically the rights of translation, reprinting, reuse of illustrations, recitation, broadcasting, reproduction on microfilms or in any other physical way, and transmission or information storage and retrieval, electronic adaptation, computer software, or by similar or dissimilar methodology now known or hereafter developed.
The use of general descriptive names, registered names, trademarks, service marks, etc. in this publication does not imply, even in the absence of a specific statement, that such names are exempt from the relevant protective laws and regulations and therefore free for general use.
The publisher, the authors and the editors are safe to assume that the advice and information in this book are believed to be true and accurate at the date of publication. Neither the publisher nor the authors or the editors give a warranty, expressed or implied, with respect to the material contained herein or for any errors or omissions that may have been made. The publisher remains neutral with regard to jurisdictional claims in published maps and institutional affiliations.

This Springer imprint is published by the registered company Springer Nature Switzerland AG
The registered company address is: Gewerbestrasse 11, 6330 Cham, Switzerland

If disposing of this product, please recycle the paper.

Preface

Integrating automated systems into daily life has become a central objective for industry and academia in the modern era. However, many automated systems that humans increasingly rely on cannot adapt to the constant evolution of the modern world. This problem, also known as drift phenomena, impacts the adoption and efficacy of automated systems in multiple critical domains (e.g., medical, finance, manufacturing, and cybersecurity) and across different data types (e.g., images and text). While there has been a recent focus on the "distribution drift," this umbrella term fails to decouple the difference between data drift and concept drift, causing misunderstandings within the research and practitioner communities. To address this gap, we proposed the DELTA workshop to bring together researchers and practitioners to uncover the relationships between types of drift and work towards a practical, reality-informed, and human-centric framework for coping with the multifaceted definitions of drift in evolving landscapes.

It is with great pleasure that we present this volume, which encompasses the proceedings of the Discovering Drift Phenomena in Evolving Landscapes (DELTA 2024) workshop.[1] The workshop was held on August 26, 2024, in conjunction with the ACM SIGKDD International Conference on Knowledge Discovery and Data Mining (KDD 2024) in Barcelona, Spain. This event served as a platform for a dynamic community of researchers and practitioners to engage with the intricate challenges of drift phenomena in complex and evolving data landscapes. The DELTA 2024 workshop featured two distinguished keynote speakers, João Gama (University of Porto, Portugal) and Heitor Murilo Gomes (Victoria University of Wellington, New Zealand), whose contributions enriched the discussions and advanced understanding of the field. Gama presented a neural-symbolic system integrating unsupervised autoencoders for anomaly detection with supervised rule learners for explanation. Murilo Gomes addressed challenges in concept drift, focusing on detection, simulation, evaluation, and handling partially and delayed labeled data in learning algorithms.

The workshop began with an opening session that introduced the overarching themes and aligned the attendees with ongoing research collaborations, specifically touching on the synergies with the Vitality Project. Following this, the first set of oral presentations (*Adaptive and Robust Learning in Dynamic Environments*) covered adaptive and robust machine-learning techniques for dynamic and constrained environments. Key topics in this session included methods for adapting machine learning models in resource-limited settings, frameworks for continual learning across shifting domains, and novel embedding techniques that leverage transformers for dynamic networks. The session closed with the invited talk by Heitor Murilo Gomes. Then, the second set of oral presentations (*Challenges and Solutions in Drift Detection and Anomaly Explanation*) explores the nuances of distribution drifts in industrial and misinformation contexts and visualizes concept drift in image streams. The session closed with the invited talk by João Gama.

---

[1] https://aiimlab.org/events/KDD_2024_Discovering_Drift_Phenomena_in_Evolving_Landscape.html

The workshop then shifted to the interactive session (*Innovative Approaches to Concept Drift Detection and Landscape Shifts*), where participants engaged in discussions on benchmarking local drift detection, unsupervised drift detection using parallel neural network activations, and novel approaches for assessing landscape shifts through topological measures. The session concluded with a focus on temporal dependencies and data stream forgetting, the assignment of the Best Paper Award (given to Shruti Saxena), and the final remarks.

This year's workshop witnessed the submission of 17 high-quality research articles. Each submission underwent a rigorous, double-blind review process, evaluated by at least three independent referees who assessed the work based on originality, scientific quality, clarity of presentation, the significance of the contribution, and transparency in providing access to code and datasets. The selection process was thorough and reflected the commitment to high academic standards, resulting in the acceptance of 9 full papers and 1 short paper after the authors made the necessary revisions.

We take this opportunity to extend our sincere appreciation to the program committee members, whose expertise spans around 20 different countries. Their dedication and diligent evaluations were essential in ensuring the quality and impact of the selected contributions. The diversity and breadth of the committee members enriched the review process, fostering a global perspective on the challenges and solutions in the field of drift phenomena. We also wish to commend the authors for their innovative and impactful contributions. Their work has provided new perspectives and advanced the discussion on the intricacies of drift phenomena, setting a strong foundation for future research. The collection of papers in this volume reflects a wide range of approaches and findings that promise to inspire further exploration and collaboration within the community. Special thanks to Springer and its staff for their invaluable support in preparing the post-proceedings and ensuring these works' timely and professional publication.

These proceedings testify to the exceptional engagement and vibrant discussions experienced by more than fifty participants during the event. They will serve as a valuable resource for researchers, practitioners, and scholars, sparking new ideas and conversations to advance this dynamic and crucial study area. We look forward to the continued growth and exploration of drift phenomena, supported by the knowledge exchange fostered at the DELTA workshop.

October 2024

Marco Piangerelli
Bardh Prenkaj
Ylenia Rotalinti
Ananya Joshi
Giovanni Stilo

Organization

General Chairs

Marco Piangerelli	University of Camerino, Italy
Bardh Prenkaj	Technical University of Munich, Germany
Ylenia Rotalinti	Brunel University, UK
Ananya Joshi	Carnegie Mellon University, USA
Giovanni Stilo	University of L'Aquila, Italy

Program Committee

Jean Paul Barddal	Pontifícia Universidade Católica do Paraná, Brazil
Dariusz Brzezinski	Poznań University of Technology, Poland
Andrea D'Angelo	University of L'Aquila, Italy
Apurba Das	BITS Pilani, Hyderabad Campus, India
Anxhelo Diko	Sapienza University of Rome, Italy
Lei Du	Northwestern Polytechnical University, China
Raul Sena Ferreira	Continental Engineering Services, Germany
Jorge Garcia-Gutierrez	University of Seville, Spain
Sandra Geisler	RWTH Aachen University, Germany
Kanu Goel	Punjab Engineering College, India
Mohammad Rez. Huq	East West University, Bangladesh
Bijay Prasad Jaysawal	National Cheng Kung University, Taiwan
Botao Jiao	China University of Petroleum, China
Ananya Joshi	Carnegie Mellon University, USA
Koki Kawabata	Osaka University, Japan
A. I. Kontaxakis	Université libre de Bruxelles, Belgium
Thomas Lacombe	University of Auckland, New Zealand
Sangmin Lee	Kwangwoon University, South Korea
Zeyi Liu	Tsinghua University, China
Jie Lu	Information Engineering University, China
Saulo M. Mastelini	University of São Paulo, Brazil
Shikha Mehta	Jaypee Institute of Information Technology, India
Abdul Sattar Palli	Universiti Teknologi PETRONAS, Malaysia
Marco Piangerelli	University of Camerino, Italy
Bardh Prenkaj	Technical University of Munich, Germany
Muhammad Rashid	University of Torino, Italy

Ylenia Rotalinti	Brunel University, UK
Negin Samadi	University of Tabriz, Iran
A. Erdem Sariyuce	University at Buffalo, USA
George Siachamis	TU Delft, The Netherlands
Paula Silva	INESC TEC - LIAAD, Portugal
Giovanni Stilo	University of L'Aquila, Italy
Mauro D. L. Tosi	Luxembourg Institute of Science and Technology, Luxembourg
Bruno Veloso	INESC TEC & FEP-UP, Portugal
Kun Wang	University of Technology Sydney, Australia & Shanghai University, China
Akihiro Yamaguchi	Toshiba Corporation, Japan

Contents

Adaptive and Robust Learning in Dynamic Environments

Adaptive Machine Learning for Resource-Constrained Environments 3
    Sebastián A. Cajas Ordóñez, Jaydeep Samanta,
    Andrés L. Suárez-Cetrulo, and Ricardo Simón Carbajo

RoCoNA: A Robust Continual Learning Framework for Alignment
of Dynamic Networks Under Distribution Shift and Domain Differences 20
    Shruti Saxena and Joydeep Chandra

CeDFormer: Community Enhanced Transformer for Dynamic Network
Embedding 37
    Jiaqi Guo, Tianpeng Li, Minglai Shao, Wenjun Wang, Lin Pan,
    Xue Chen, and Yueheng Sun

Challenges and Solutions in Drift Detection and Anomaly Explanation

On the Impact of Industrial Delays when Mitigating Distribution Drifts:
An Empirical Study on Real-World Financial Systems 57
    Thibault Simonetto, Maxime Cordy, Salah Ghamizi, Yves Le Traon,
    Clément Lefebvre, Andrey Boystov, and Anne Goujon

Understanding Knowledge Drift in LLMs Through Misinformation 74
    Alina Fastowski and Gjergji Kasneci

Exploring Concept Drift Visualization and Explanation in Image Streams 86
    Giacomo Ziffer and Emanuele Della Valle

Innovative Approaches to Concept Drift Detection and Landscape Shifts

A Synthetic Benchmark to Explore Limitations of Localized Drift
Detections 101
    Flavio Giobergia, Eliana Pastor, Luca de Alfaro, and Elena Baralis

Unsupervised Concept Drift Detection Based on Parallel Activations
of Neural Network 111
    Joanna Komorniczak and Paweł Ksieniewicz

Unsupervised Assessment of Landscape Shifts Based on Persistent
Entropy and Topological Preservation 128
  Sebastián Basterrech

Addressing Temporal Dependence, Concept Drifts, and Forgetting in Data
Streams 146
  Federico Giannini and Emanuele Della Valle

Author Index 165

Adaptive and Robust Learning in Dynamic Environments

Adaptive Machine Learning for Resource-Constrained Environments

Sebastián A. Cajas Ordóñez, Jaydeep Samanta,
Andrés L. Suárez-Cetrulo(✉), and Ricardo Simón Carbajo

Ireland's Centre for Applied Artificial Intelligence (CeADAR),
University College Dublin, Dublin, Ireland
{sebastian.cajasordonez,jaydeep.samanta,andres.suarez-cetrulo,
ricardo.simoncarbajo}@ucd.ie

Abstract. The Internet of Things is an example domain where data is perpetually generated in ever-increasing quantities, reflecting the proliferation of connected devices and the formation of continuous data streams over time. Consequently, the demand for ad-hoc, cost-effective machine learning solutions must adapt to this evolving data influx. This study tackles the task of offloading in small gateways, exacerbated by their dynamic availability over time. An approach leveraging CPU utilization metrics using online and continual machine learning techniques is proposed to predict gateway availability. These methods are compared to popular machine learning algorithms and a recent time-series foundation model, Lag-Llama, for fine-tuned and zero-shot setups. Their performance is benchmarked on a dataset of CPU utilization measurements over time from an IoT gateway and focuses on model metrics such as prediction errors, training and inference times, and memory consumption. Our primary objective is to study new efficient ways to predict CPU performance in IoT environments. Across various scenarios, our findings highlight that ensemble and online methods offer promising results for this task in terms of accuracy while maintaining a low resource footprint. Code is available at https://github.com/sebasmos/AML4CPU.

Keywords: Data streams · Machine learning · IoT · Edge Computing

1 Introduction

In today's dynamic world, data streams are abundant and ever-changing, particularly in non-stationary environments where data undergoes constant change. Machine learning (ML) models in these domains may need regular model updates to minimize the degradation of their performance over time, as seen in weather prediction and customer preference model [1], social networks, sensor networks, and financial data streams [2]. The growing volume of data generated underscores a pressing need for real-time processing capabilities. This perpetual evolution in data undermines model predictions, as outdated data distributions may no longer align with current data, necessitating frequent model updates and introducing the challenge of concept drift in data streams [1]. Such limitations can impede AI systems, rendering them incapable of effectively adapting to ongoing changes and struggling with memory constraints to process incoming data [3]. Furthermore, the evolving dynamics of the data streams, whereby behaviors may also re-occur, make it necessary to consider forgetting mechanisms [4] and reusing former active learners [2].

CPU demand is a primary driver of resource shortages in virtualized environments, significantly impacting host-machine performance [5,6]. Accurately predicting future resource usage for impending demands stands as one of the significant challenges in cloud computing [7], which is particularly challenging due to the non-stationarity of CPU utilization and the potential presence of concept drifts [8]. This can result in inefficient resource allocation across machines. Thus, forecasting CPU allocation accurately can help reduce energy consumption [9]. Such non-stationarity may arise from many background processes tracing periodic and non-periodic behavior with sudden peaks of loads [10]. Hence, estimating CPU utilization levels can be crucial in aligning tasks with resources, maximizing their availability, and minimizing computational costs [11].

Traditionally, statistical predictive models such as Autoregressive Integrated Moving Average (ARIMA) and family variations have focused on optimizing cost functions [12], resulting in a good fit to the data but limiting their adaptability for non-linear-trends and long-term dependencies [13,14]. More recently, neural networks have shown stronger capabilities to fill that gap; for example, one-dimensional Convolutional Neural Networks (CNN) have shown their effectiveness for pattern extraction on 1-dimensional complex signals [15] and nonlinearity time-series extraction [16]. Similarly, as introduced Long Short-Term Memory (LSTM) networks [17] to cover the vanishing gradient problem of Recurrent Neural Networks (RNN) which has allowed neural networks to learn longer-term dependencies that have extended over all LSTM-based models [10,11] outperforming traditional methods [18]. Finally, online incremental ML algorithms allow drift handling in data streams with an efficiency that suits resource-aware environments [2]. This allows for quick adjustments to temporal changes, largely owing to the incorporation of forgetting mechanisms, ensuring rapid adaptation to new patterns [4].

This paper aims to be a comparative study of the performance of the online regression models on a novel application dataset of CPU loads that exhibits non-stationary patterns over time. Our study delves into a comparative analysis of various classic ML models alongside online learning algorithms, assessing accuracy and computation performance metrics. Furthermore, these models are juxtaposed with recent deep learning methods and the time-series foundation model Lag-Llamma [19]. This research aims to contribute to the advancement of IoT systems by providing insights into model selection for CPU performance estimation through the use of online ML and other modern algorithms. This work also provides insights into the suitability of online regression models in our application domain. The main contributions of this paper are outlined below.

1. *CPU utilization prediction*: This paper proposes an approach to predicting CPU load in IoT gateways using state-of-the-art ML algorithms. While offline ensemble methods offer the best trade-off between accuracy and computational cost, continual learning methods offer promising results in predicting CPU loads accurately for edge devices.
2. *Evaluation benchmark:* A benchmark is proposed to compare traditional versus online and foundation models. In addition to performance evaluation, the memory and runtime of the models are computed as a measure of their footprint. This assessment allows the identification of the most effective model for CPU performance estimation and considers the chosen approach's computational and environmental implications.
3. *Code and data sharing*: The code and data generation used in our experiments are publicly available to facilitate reproducible research and encourage collaboration in the research community.

The paper is structured as follows: After the introduction in Sect. 1, Sect. 2 covers related work on CPU utilization prediction and the techniques utilized in this paper. Next, Sect. 3 describes the proposed dataset. Subsequently, Sect. 4 presents the experiments, outlining the metrics and models used. Section 5 includes a discussion of the results obtained, and Sect. 6 presents the conclusion.

2 Related Work

Predicting CPU utilization has been approached through different methods, including more traditional methods such as polynomial fitting [5], regression-based models such as linear regression [20], and gradient-descent optimizers like stochastic gradient descend (SGD) [21]. Other advanced methods include adaptive networks with clustering [6] and stack generalization, which combines algorithms such as KNN and decision trees (DT).

Shaikh et al. [22] used DTs [23] to forecast CPU usage in VM workloads. DTs start learning by splitting the first node based on a metric such as information gain or the Gini coefficient [24]. This split triggers the creation of new nodes, which may split again during the learning process. Final nodes without children predict outcomes for both classification and regression tasks.

Based on base models such as DTs, ensemble methods can be constructed by aggregating multiple predictive models (weak learners) for improved accuracy, relying on a voting mechanism. These have recently been used for VM resource allocation [25]. Some example methods are Adaboost, XGBoost [26], and random forests (RF). RF [27] specifically builds upon DTs by training multiple decision trees on different data subsets to promote diversity and make predictions based on the majority vote, enhancing accuracy and stability.

Recent work on predicting data center workloads includes Kim et al.'s study [28], which combines Linear Regression, support vector machines, and time-series models with dynamic weight adjustment. Additionally, support vector regression [29] and Kalman smoothing [30] have demonstrated effectiveness in handling dynamic characteristics for accurate predictions on CPU load and cloud prediction. Another incremental approach worth exploring for this task is the Passive-Aggressive algorithm (PA) [31], which adapts the model based on feedback and can help with the dynamic nature of the CPU load changes. This was initially proposed for binary classification, incrementally updating the decision boundary, and later extended to regression tasks.

Neural-network-based approaches [7–10], LSTMs [17], and hybrid models combining ensembling models with LSTMs [11] have also recently been used for CPU utilization prediction. LSTMs are a type of RNN designed to address the vanishing gradient problem that affects standard RNNs and as potent tools for processing and forecasting time series data across diverse domains [7,32]. Mason et al. (2018) specifically explored the potential of neural networks in CPU utilization forecasting, developing evolutionary neural networks through an evolutionary optimization algorithm. Moreover, LSTMs have been employed in CPU utilization forecasting, often compared against traditional techniques like ARIMA [9]. Additionally, various architectures such as Recurrent Neural Networks (RNNs), Bidirectional LSTMs (BiLSTMs), and hybrid versions like BiLSTM-RNN models and CNN-LSTM [14] have demonstrated applicability in this domain [10,14].

Hoeffding Trees (HT) [33] were originally designed for constructing and updating decision trees in dynamic data streams. Leveraging the Hoeffding bound, a statistical inequity, they efficiently determine optimal splits at each node without requiring full dataset analysis, thus becoming highly memory-efficient. The Hoeffding Adaptive Tree (HAT) enhances adaptiveness by replacing old branches dynamically using metrics such as Adaptive Windowing (ADWIN) algorithm [34] and also proposes a bootstrapping sampling on top of Hoeffding Trees. Bagging [35] and boosting-based [36] techniques have recently proven their success as part of ensembles in data stream learning like Adaptive Random Forests (ARF) [3] and Streaming Random Patches (SRP) [37]. ARF [3] is an enhanced adaptive ensemble with diversity through resampling and random node splitting, equipped with drift detection per node for adaptive training. ARF uses enhanced HTs as base learners and ADWIN as a drift detector to understand when to train and replace decision trees.

Finally, the advent of foundation models in artificial intelligence has created a trend for reusing pre-trained models, something already common in the data stream learning field [2]. Lag-Llama has recently emerged as a time-series foundational model [19], leveraging the properties of the decoder-only transformer-based architecture LLaMA and incorporating pre-normalization via the RMSNorm . A current topic of discussion in foundation models, which tend to be multi-purpose and thus experiment domain drifts over time, is the issue of alignment and models sharing a similar world representation. This topic has an analogy in data stream learning, as models representing similar data distributions are often contrasted with each other in the meta-learning field, comparing the similarity of the data fed to them (*concept similarity*) or their predictive results (*conceptual equivalence*) [2].

Predicting CPU performance efficiently is crucial for optimizing system resources and enhancing overall computational efficiency. In this comparative study, we delve into the performance evaluation of state-of-the-art classical models, deep learning, online ML, and a time-series foundational in both zero-shot and fine-tuned setups for this predictive task. Our research endeavors to contribute to advancing new methodologies for CPU performance estimation tasks and offer a new dataset for data stream learning.

3 Research Data

The hardware utilized for data collection was an Orange Pi 5[1], powered by the 8-core RK3588S processor.

The data collection was performed using the *psutil* library, recording CPU usage per core and UNIX timestamps at 1-minute intervals. This process ran over ≈ 32 days (47,315 min) while subjecting the system to stress tests using the *stress-ng* tool[2]. *Stress-ng* simulates diverse workloads, engaging all CPU cores at varying utilization levels (0–100%) through random generation. This used workloads of 60 min followed by a 60-second pause before initiating the next test. To isolate CPU behavior, the *stress-ng test* was configured to focus exclusively on CPU usage.

The collected samples underwent a resampling process to ensure an evenly distributed index with precisely one-minute intervals between each sample. The 47,315 data samples were partitioned into 37,852 samples (80%) for training and 9,463 (20%) for testing. The datasets used in this paper are publicly available in the data folder of our GitHub repository[3]. This includes training and testing sets, named *train_data.csv* and *test_data.csv*, respectively.

The feature set used in the experiments is univariate, using lags of CPU utilization to predict the next one; thus, models in this paper will provide 1-minute ahead predictions. Different experiments have used different window sizes (WS) or lags lengths L, where L refers to the past CPU utilization measurements $x(t-1)$, $x(t-7)$, $x(t-14)$,..., $x(t-L)$, where L is the maximum lag index used in the model. The target feature is the CPU utilization one step ahead, $x(t)$.

4 Experiments

Three distinct experiments were carried out to evaluate the optimal models using the CPU dataset over 20 different seeds.

- *Experiment I*: A hold-out benchmarking process was conducted between state-of-the-art ML algorithms.

---

[1] http://www.orangepi.org/html/hardWare/computerAndMicrocontrollers/details/Orange-Pi-5.html.
[2] https://wiki.ubuntu.com/Kernel/Reference/stress-ng.
[3] https://github.com/sebasmos/AML4CPU/tree/main/data.

- *Experiment II*: Online incremental learners were evaluated using the training and test sets from Experiment I for pre-training and for a prequential evaluation [2] respectively.
- *Experiment III*: A zero-shot and fine-tuning setup of the time-series foundation model Lag-Llama was run as in the previous experiments to compare the generalization capabilities of foundation models against other state-of-the-art and online ML methods.

Experiments 1 and 2 were conducted using the default hyper-parameter settings for each library, while only adjusting the *window size* parameter-related to the length of the feature set. Experiment 3 consisted of two parts: the first focusing on zero-shot and the second on fine-tuning, both using the default settings for lag-llama, including context length and RoPE.

Each experiment was run 20 times with different seeds to handle non-deterministic models, providing mean and standard deviation across runs and boxplots for them. The libraries used for this study were *river* for online ML, *scikit-learn* for classical methods, and *PyTorch* for deep learning. Detailed results are provided in Tables 1, 2, and 3. In these tables, we highlight the best results for each algorithm across different WSs marked in bold. The overall best results are further emphasized with an underline, and we will focus our analysis on these bold and underlined results. Boxplots and scatterplots exhibiting similar patterns in this experimental section are also excluded to simplify the analysis.

To assess model performance, this work employed a variety of error metrics [38]. These are covered below with their mathematical intuition. N represents the number of data points in the dataset, y_i represents the actual value, and $\hat{y}i$ represents the predicted value of ith data point in the dataset.

- *Mean Absolute Error (MAE)* is computed as the average of the absolute difference between the predicted and actual values: $MAE(y, \hat{y}) = \frac{\sum_{i=0}^{N-1} |y_i - \hat{y}_i|}{N}$.
- *Mean Squared Error (MSE)* measures the average squared difference between the actual and predicted values: $MSE(y, \hat{y}) = \frac{\sum_{i=0}^{N-1} (y_i - \hat{y}_i)^2}{N}$
- *Root Mean Squared Error (RMSE)* is the square root of the MSE and can be represented as $RMSE(y, \hat{y}) = \sqrt{\frac{\sum_{i=0}^{N-1} (y_i - \hat{y}_i)^2}{N}}$.
- *Mean Absolute Percentage Error (MAPE)* measures the average absolute percentage difference between the actual and predicted values: $MAPE(y, \hat{y}) = \frac{100\%}{N} \sum_{i=0}^{N-1} |\frac{y_i - \hat{y}_i}{y_i}|$.
- Symmetric mean absolute percentage error (SMAPE) is introduced to overcome the asymmetric nature of MAPE: $SMAPE(y, \hat{y}) = \frac{100\%}{N} \sum_{i=0}^{N-1} \frac{2*|y_i - \hat{y}_i|}{|y| + |\hat{y}|}$.
- *Mean Absolute Scaled Error (MASE)* is determined by calculating the mean absolute error of actual forecasts and the mean absolute error produced by a naive forecast calculated using the in-sample data. $MASE = \frac{MAE}{MAE_{in-sample, naive}}$

- *R-squared error* (R^2) measures the percentage of the target variable's overall variance that can be accounted for by the model's predictions: $R^2(y, \hat{y}) = 1 - \frac{\sum_{i=1}^{N}(y_i - \hat{y}_i)^2}{\sum_{i=1}^{N}(y_i - \bar{y})^2}$.

In addition, measurements of training, evaluation time, and memory consumption per model using *asizeof.asizeof(model)* were captured to understand the model's footprints. Training and evaluation times were measured in seconds, and memory was measured in megabytes (MB). These experiments have been run in a server with 32-core AMD Ryzen Threadripper PRO 5975WX, 256 GB of RAM, and 2 x NVIDIA GeForce RTX 4090 GPUs. The GPU has mainly been used for Lag-LLama in Experiment III, while the rest of the algorithms have been run in CPU to allow comparable runtimes.

4.1 Experiment I

Firstly, state-of-the-art ML models are compared as detailed in Table 1. Subsequently, we evaluate their performance using hold-out validation. The outcomes are visually represented through model boxplots and scatterplots in Figs. 1 and 2.

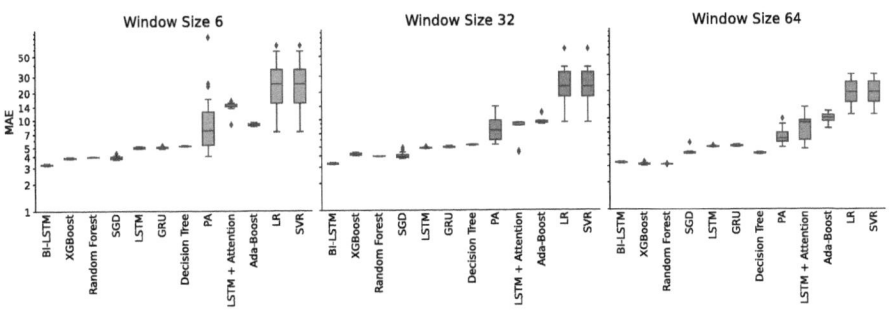

Fig. 1. MAE per model in Experiment I at different window sizes.

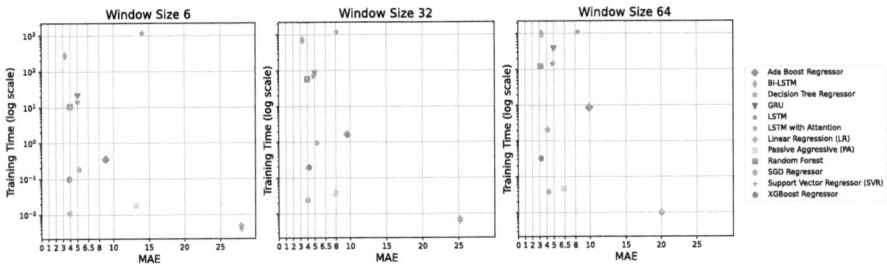

Fig. 2. Training time vs. MAE per model in Experiment I.

Table 1. Experiment I with highlighted results for WS with the lowest MAE across 20 runs. Values are rounded to a maximum of three decimal places.

Model	WS	MAE		RMSE		SMAPE		R^2		MASE		Train time (s)	Evaluation (s)	Memory
		mean	std	mean	std	mean	std	mean	std	mean	std	mean	mean	mean
XGBoost Regressor	6	3.845	0.056	9.391	0.065	22.229	0.318	0.903	0.001	0.994	0.014	0.099	0.003	0.004
	9	3.902	0.079	9.408	0.082	22.231	0.439	0.902	0.002	1.008	0.02	0.11	0.003	0.004
	12	3.963	0.089	9.472	0.095	22.413	0.544	0.901	0.002	1.024	0.023	0.123	0.002	0.004
	20	4.045	0.107	9.474	0.104	23.037	0.493	0.901	0.002	1.044	0.028	0.158	0.003	0.004
	32	4.178	0.111	9.553	0.113	23.329	0.468	0.899	0.002	1.078	0.029	0.198	0.013	0.004
	64	**3.185**	**0.103**	**7.344**	**0.424**	**21.881**	**0.381**	**0.941**	**0.007**	**0.822**	**0.027**	**0.322**	**0.01**	**0.004**
Ada Boost Regressor	6	8.933	0.23	12.358	0.244	32.739	0.626	0.831	0.007	2.308	0.06	0.35	0.003	0.013
	9	**8.901**	**0.238**	**12.381**	**0.3**	**32.699**	**0.637**	**0.83**	**0.008**	**2.303**	**0.061**	**0.449**	**0.003**	**0.013**
	12	9.266	0.189	12.969	0.175	33.432	0.48	0.814	0.005	2.393	0.049	0.745	0.004	0.015
	20	10.32	1.814	14.282	2.191	35.313	3.275	0.77	0.075	2.664	0.468	1.154	0.005	0.015
	32	9.609	0.974	13.713	1.185	34.35	1.793	0.791	0.029	2.479	0.251	1.685	0.005	0.014
	64	9.765	1.114	12.932	1.009	36.262	2.16	0.815	0.029	2.521	0.288	8.665	0.021	0.04
Decision Tree Regressor	6	5.221	0.033	13.468	0.105	29.861	0.261	0.8	0.003	1.349	0.008	0.184	0.003	0.002
	9	5.289	0.047	13.621	0.14	29.54	0.168	0.795	0.004	1.366	0.012	0.272	0.003	0.002
	12	5.271	0.052	13.702	0.169	28.972	0.184	0.793	0.005	1.361	0.014	0.362	0.003	0.002
	20	5.194	0.05	13.383	0.136	28.793	0.359	0.802	0.004	1.341	0.013	0.599	0.003	0.002
	32	5.254	0.045	13.407	0.131	30.717	0.195	0.802	0.004	1.355	0.012	0.966	0.003	0.002
	64	**4.119**	**0.074**	**9.783**	**0.294**	**26.17**	**0.177**	**0.895**	**0.006**	**1.065**	**0.019**	**2.065**	**0.003**	**0.003**
Random Forest Regressor	6	3.967	0.012	9.407	0.014	22.216	0.11	0.902	0.001	1.025	0.003	10.811	0.168	0.086
	9	3.932	0.013	9.308	0.021	21.56	0.11	0.904	0.001	1.016	0.003	16.121	0.169	0.083
	12	3.924	0.012	9.281	0.021	21.33	0.127	0.905	0.001	1.014	0.003	21.552	0.168	0.083
	20	3.907	0.011	9.269	0.02	21.152	0.11	0.905	0.001	1.009	0.003	36.062	0.168	0.083
	32	3.95	0.012	9.348	0.032	21.555	0.129	0.904	0.001	1.019	0.003	58.982	0.169	0.083
	64	**3.142**	**0.02**	**7.525**	**0.087**	**20.195**	**0.094**	**0.938**	**0.001**	**0.811**	**0.005**	**121.804**	**0.164**	**0.085**
Passive Aggressive Regressor	6	13.171	17.324	18.766	18.929	45.013	29.659	0.235	2.193	3.403	4.476	0.017	0.003	0.004
	9	14.211	19.874	20.498	21.539	46.119	30.501	0.049	2.434	3.671	5.134	0.019	0.007	0.004
	12	12.34	11.747	17.91	12.425	46.945	24.475	0.484	0.836	3.187	3.034	0.023	0.005	0.004
	20	7.659	3.185	12.663	2.617	34.028	8.427	0.816	0.083	1.977	0.822	0.027	0.006	0.004
	32	7.717	2.278	12.804	1.952	36.309	9.809	0.815	0.061	1.991	0.588	0.035	0.009	0.004
	64	**6.38**	**1.379**	**9.729**	**1.228**	**34.621**	**5.144**	**0.894**	**0.028**	**1.646**	**0.357**	**0.046**	**0.002**	**0.005**
SGD Regressor	6	3.913	0.148	9.8	0.015	22.547	0.616	0.894	0.001	1.011	0.038	0.012	0.003	0.004
	9	3.997	0.215	9.81	0.034	22.908	0.718	0.894	0.001	1.033	0.056	0.017	0.007	0.004
	12	3.946	0.148	9.806	0.015	22.789	0.671	0.894	0.001	1.019	0.038	0.018	0.004	0.004
	20	**3.886**	**0.203**	**9.817**	**0.02**	**22.288**	**0.977**	**0.894**	**0.001**	**1.003**	**0.053**	**0.019**	**0.001**	**0.004**
	32	4.051	0.327	9.839	0.059	22.971	1.147	0.893	0.001	1.045	0.084	0.026	0.005	0.004
	64	4.236	0.286	7.556	0.154	25.344	0.764	0.937	0.003	1.094	0.074	0.041	0.007	0.005
LSTM	6	5.001	0.104	11.007	0.07	25.666	0.284	0.866	0.002	1.292	0.027	14.475	0.006	0.066
	9	4.978	0.094	10.926	0.093	25.42	0.327	0.868	0.002	1.286	0.024	21.487	0.01	0.066
	12	**4.811**	**0.098**	**10.765**	**0.081**	**24.574**	**0.411**	**0.872**	**0.002**	**1.243**	**0.025**	**28.351**	**0.013**	**0.066**
	20	4.865	0.101	10.694	0.082	24.566	0.449	0.874	0.002	1.256	0.026	45.884	0.045	0.066
	32	4.855	0.086	10.679	0.084	24.726	0.329	0.874	0.002	1.252	0.022	70.148	0.069	0.066
	64	4.853	0.084	10.675	0.086	24.777	0.334	0.875	0.002	1.253	0.022	166.131	0.18	0.066
BI-LSTM	6	**3.279**	**0.043**	**7.448**	**0.07**	**19.899**	**0.592**	**0.939**	**0.001**	**0.847**	**0.011**	**279.032**	**0.178**	**0.131**
	9	3.279	0.043	7.449	0.071	19.905	0.595	0.939	0.001	0.847	0.011	359.11	0.259	0.131
	12	3.28	0.043	7.45	0.071	19.911	0.599	0.939	0.001	0.847	0.011	408.813	0.257	0.131
	20	3.283	0.043	7.456	0.07	19.928	0.593	0.939	0.001	0.848	0.011	529.375	0.559	0.131
	32	3.286	0.043	7.461	0.07	19.952	0.592	0.939	0.001	0.848	0.011	716.442	0.774	0.131
	64	3.291	0.044	7.462	0.073	19.994	0.604	0.939	0.001	0.85	0.011	977.366	1.052	0.131
Gated Recurrent Units	6	5.047	0.117	10.71	0.12	25.807	0.264	0.873	0.003	1.304	0.03	32.276	0.028	0.049
	9	5.005	0.092	10.633	0.131	25.722	0.204	0.875	0.003	1.293	0.024	84.692	0.052	0.049
	12	5.004	0.107	10.55	0.107	25.775	0.233	0.877	0.003	1.292	0.028	102.017	0.045	0.049
	20	**4.961**	**0.097**	**10.497**	**0.09**	**25.648**	**0.219**	**0.878**	**0.002**	**1.281**	**0.025**	**76.746**	**0.037**	**0.049**
	32	4.978	0.091	10.507	0.091	25.793	0.241	0.878	0.002	1.284	0.023	108.012	0.09	0.049
	64	4.981	0.094	10.507	0.1	25.871	0.173	0.879	0.002	1.286	0.024	362.131	0.123	0.049
LSTM with Attention	6	13.373	3.154	19.175	3.599	42.255	6.426	0.581	0.135	3.455	0.815	1138.691	0.414	0.213
	9	9.537	4.483	15.36	4.769	34.566	9.085	0.716	0.169	2.464	1.158	1112.52	0.413	0.216
	12	8.46	4.342	13.988	4.717	32.983	9.179	0.761	0.172	2.185	1.122	1045.146	0.411	0.219
	20	**7.258**	**3.805**	**12.622**	**3.893**	**30.334**	**7.686**	**0.809**	**0.137**	**1.874**	**0.982**	**1083.556**	**0.423**	**0.227**
	32	8.248	2.427	12.791	1.997	33.031	4.521	0.816	0.062	2.128	0.626	1122.657	0.41	0.239
	64	7.94	2.849	11.637	2.633	32.214	5.739	0.844	0.076	2.05	0.736	1052.156	0.397	0.27

(*continued*)

Table 1. (*continued*)

Model	WS	MAE		RMSE		SMAPE		R^2		MASE		Train time (s)	Evaluation (s)	Memory
		mean	std	mean	std	mean	std	mean	std	mean	std	mean	mean	mean
Linear Regression	6	28.025	15.613	33.05	17.468	65.17	22.007	-0.527	1.661	7.241	4.034	0.005	0.001	0.001
	9	25.464	9.304	30.238	10.079	61.45	14.192	-0.116	0.737	6.579	2.404	0.005	0.001	0.001
	12	25.337	9.129	29.865	9.816	62.513	14.505	-0.086	0.731	6.544	2.358	0.005	0.001	0.001
	20	24.673	11.559	29.545	12.59	61.166	18.031	-0.129	1.065	6.37	2.984	0.005	0.001	0.001
	32	25.19	12.355	30.552	13.313	61.847	19.29	-0.214	1.16	6.498	3.187	0.006	0.001	0.001
	64	**20.05**	**5.906**	**24.994**	**6.018**	**54.479**	**9.883**	**0.276**	**0.344**	**5.177**	**1.525**	**0.009**	**0.001**	**0.001**
SVR	6	28.025	15.613	33.05	17.468	65.17	22.007	-0.527	1.661	7.241	4.034	0.005	0.001	0.001
	9	25.464	9.304	30.238	10.079	61.45	14.192	-0.116	0.737	6.579	2.404	0.005	0.001	0.001
	12	25.337	9.129	29.865	9.816	62.513	14.505	-0.086	0.731	6.544	2.358	0.005	0.001	0.001
	20	24.673	11.559	29.545	12.59	61.166	18.031	-0.129	1.065	6.37	2.984	0.005	0.001	0.001
	32	25.19	12.355	30.552	13.313	61.847	19.29	-0.214	1.16	6.498	3.187	0.006	0.001	0.001
	64	**20.05**	**5.906**	**24.994**	**6.018**	**54.479**	**9.883**	**0.276**	**0.344**	**5.177**	**1.525**	0.005	**0.001**	**0.001**

Boxplots for WS 6, 9, and 12 behave similarly in terms of MAE. Consequently, only boxplots for WSs 6, 32, and 64 are then presented in Fig. 1. The interquartile range (IQR) can define a consistent MAE, which helps assess model stability and variability.

XGBoost and Random Forest consistently perform well across all window sizes (WSs), demonstrating low interquartile ranges (IQRs) and stable error rates. Similarly, BI-LSTM performs well but at a higher training and inference time and memory consumption. Larger window sizes tend to enhance stability in error rates across most models.

When the MAE is low and the RMSE is high in Table 1, it indicates that while the average error is small, there are occasional large errors that cause the RMSE to be large. This is evident in models like support vector regression (SVR), linear regression (LR), and LSTM with attention. Indeed, LR and SVR are by far the worst-performing methods compared across all predictive error metrics. Conversely, models with very low RMSE, such as XGBoost, Adaboost, the default *scikit-learn's* decision tree (CART), random forest (RF), stochastic gradient descent (SGD), LSTM, and BI-LSTM, indicate that their predictions do not tend to have spikes with large deviations from the ground truth. Hence being more reliable over time.

Figure 2 depicts MAE obtained by models per WS over training times. Analyzing the top five best models in terms of MAE and training times, the SGD consistently achieves the shortest training time across all WSs. XGB offers comparable results to RF in terms of MAE and boasts a faster training and inference time across all WSs. LR and SVR are the fastest models overall during training time but with a high MAE. All of the algorithms ran in this experiment have low memory consumption and inference runtime footprints, thus being suitable for edge devices.

The best overall models are XGBoost and RF, considering all factors: lowest errors, minimal training and inference times, and efficient memory usage. XGBoost offers the best balance between performance and resource consumption if training times are considered, making it ideal for resource-constrained applications that may need re-training at the edge, where low error rates and efficient use of time and memory are crucial.

SGD is the quickest model for all window sizes. While RF, XGBoost, and Bidirectional LSTM achieve minimal predictive error, the last one requires high training times in CPU, making it suitable for edge devices with in-built GPU or in cases where timely on-device re-training is not critical.

4.2 Experiment II

The second experiment evaluates online learning algorithms, as outlined in Table 2, using a *prequential evaluation*. Online ML models are envisioned to learn on the fly, continuously adapting as new data arrives. Thus, during the evaluation of this experiment, we perform model updates [2]. For more information about this process, refer to the source code in GitHub.

The algorithm with the best accuracies and lower training times in Experiment I was also evaluated prequentially in this experiment. Such evaluation entails continuous re-training, prequentially, after receiving each new data sample simulating a data stream for the XGBoost regressor, as this is not its incremental implementation. This is reflected in its evaluation time for Experiment II. The adaptive random forest algorithm and the two online DT algorithms (HAT and HT) also obtain low MAE values at WS 6, but exhibit higher memory usage and evaluation times. Despite its low pretraining time, XGBoost does not scale when being continuously retrained.

Table 2 and Fig. 3 show that XGBoosts overperforms all online learners in predictive accuracy.

In (prequential) evaluation time, various models offer a good trade-off between MAE and efficiency (see Fig. 4). Initially, HT excels in pre-training, evaluation, and memory usage for a window size 6. However, PA and SGD take the lead for larger window sizes 32 and 64. ARF obtained the second-best results in this experiment, although it underperformed offline ensembles in Experiment I.

Table 2. Results for Experiment II, showcasing the best-performing model metrics WS with the lowest MAE across 20 runs. Values are rounded to a maximum of three decimal places.

Model	Window Size	MAE		RMSE		SMAPE		R^2		MASE		Pretraining (s)	Evaluation (s)	Memory
		mean	std	mean	std	mean	std	mean	std	mean	std	mean	mean	mean
ARF	**6**	**3.427**	**0.018**	**9.078**	**0.023**	**20.17**	**0.131**	**0.909**	**0.0**	**0.885**	**0.005**	**71.431**	**31.641**	**146.031**
	9	3.553	0.038	9.212	0.066	20.291	0.162	0.906	0.001	0.918	0.01	95.346	40.418	184.937
	12	3.729	0.08	9.429	0.141	20.721	0.204	0.902	0.003	0.963	0.021	99.159	41.007	187.451
	20	4.193	0.177	9.952	0.256	21.949	0.445	0.891	0.006	1.083	0.046	113.908	42.076	157.025
	32	5.63	0.765	11.28	0.756	25.978	1.812	0.859	0.019	1.452	0.197	140.108	44.412	93.727
	64	11.661	0.213	17.159	0.24	38.946	0.481	0.676	0.009	3.011	0.055	153.104	38.984	3.986
HAT Regressor	6	3.795	0.063	9.34	0.085	21.583	0.42	0.904	0.002	0.981	0.016	4.567	2.575	2.819
	9	3.924	0.062	9.454	0.108	22.128	0.311	0.901	0.002	1.014	0.016	5.786	2.886	4.411
	12	4.039	0.085	9.61	0.203	22.626	0.624	0.898	0.004	1.043	0.022	7.075	3.185	5.815
	20	4.27	0.134	9.679	0.183	23.467	0.609	0.897	0.004	1.102	0.035	10.8	4.007	10.013
	32	6.198	3.352	11.59	3.85	28.057	6.95	0.836	0.134	1.599	0.865	17.498	5.569	10.367
	64	9.533	3.835	14.139	4.543	35.126	7.241	0.759	0.162	2.462	0.99	41.728	11.697	8.146
HT	6	3.75	0.0	9.233	0.0	20.943	0.0	0.906	0.0	0.969	0.0	3.208	2.348	2.012
	9	3.825	0.0	9.292	0.0	21.185	0.0	0.905	0.0	0.988	0.0	4.593	2.666	3.351
	12	3.894	0.0	9.337	0.0	21.637	0.0	0.904	0.0	1.006	0.0	5.935	2.948	4.851
	20	4.2	0.0	9.641	0.0	22.426	0.0	0.897	0.0	1.084	0.0	10.323	4.239	8.457
	32	4.312	0.0	9.709	0.0	23.301	0.0	0.896	0.0	1.112	0.0	17.331	6.087	12.998
	64	5.281	0.0	8.826	0.0	27.699	0.0	0.914	0.0	1.364	0.0	34.868	10.856	19.687
SRP Regressor	6	4.574	0.016	10.772	0.028	24.048	0.106	0.872	0.001	1.182	0.004	117.284	30.116	0.36
	9	4.674	0.02	10.786	0.033	24.403	0.104	0.872	0.001	1.208	0.005	135.235	34.524	0.408
	12	4.728	0.017	10.778	0.027	24.999	0.134	0.872	0.001	1.221	0.005	170.629	42.911	0.498
	20	4.759	0.024	10.772	0.034	25.663	0.165	0.872	0.001	1.229	0.006	258.696	63.778	0.797
	32	4.724	0.022	10.728	0.038	26.106	0.151	0.873	0.001	1.219	0.006	386.009	93.432	1.055
	64	5.609	0.125	11.96	0.244	28.826	0.395	0.843	0.006	1.448	0.032	718.221	175.761	1.43
PA	6	8.987	0.054	18.224	0.053	38.455	0.154	0.633	0.002	2.322	0.014	0.002	3.747	0.003
	9	8.927	0.035	17.781	0.036	37.917	0.127	0.651	0.001	2.306	0.009	0.003	3.759	0.004
	12	8.892	0.061	17.617	0.12	37.679	0.265	0.657	0.005	2.297	0.016	0.003	3.748	0.004
	20	9.501	0.056	17.651	0.089	38.875	0.188	0.656	0.003	2.453	0.015	0.004	3.755	0.004
	32	10.13	0.057	17.684	0.07	40.04	0.172	0.655	0.003	2.613	0.015	0.005	3.768	0.004
	64	6.764	0.017	10.395	0.014	34.72	0.059	0.881	0.0	1.747	0.004	0.01	3.739	0.004
SGD Regressor	6	3.897	0.008	9.814	0.001	21.661	0.007	0.894	0.0	1.007	0.002	0.002	3.73	0.004
	9	3.906	0.009	9.824	0.001	21.616	0.016	0.893	0.0	1.009	0.002	0.003	3.739	0.004
	12	3.911	0.006	9.829	0.001	21.742	0.082	0.893	0.0	1.01	0.002	0.003	3.737	0.004
	20	3.951	0.006	9.846	0.001	22.15	0.041	0.893	0.0	1.02	0.002	0.004	3.735	0.004
	32	4.015	0.012	9.861	0.002	22.494	0.036	0.893	0.0	1.036	0.003	0.005	3.743	0.004
	64	4.201	0.01	7.7		0.003	25.323	0.036	0.935	0.0	1.085	0.003	0.01	3.717	0.004
XGB regressor	6	3.846	0.066	9.393	0.081	22.302	0.298	0.903	0.002	0.994	0.017	0.157	1173.884	0.004
	9	3.898	0.076	9.41	0.071	22.279	0.319	0.902	0.001	1.007	0.02	0.143	1333.355	0.004
	12	3.956	0.094	9.452	0.097	22.345	0.431	0.901	0.002	1.022	0.024	0.167	1481.934	0.004
	20	4.032	0.103	9.462	0.094	22.788	0.532	0.901	0.002	1.041	0.027	0.223	1863.733	0.004
	32	4.157	0.095	9.53	0.078	23.327	0.499	0.9	0.002	1.072	0.025	0.261	2411.73	0.004
	64	**3.057**	**0.066**	**6.766**	**0.18**	**21.826**	**0.422**	**0.95**	**0.003**	**0.789**	**0.017**	**0.43**	**4271.722**	**0.004**

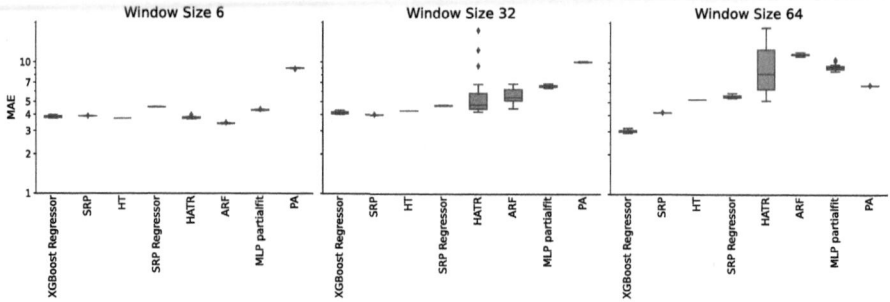

Fig. 3. MAE per model in Experiment II at different window sizes.

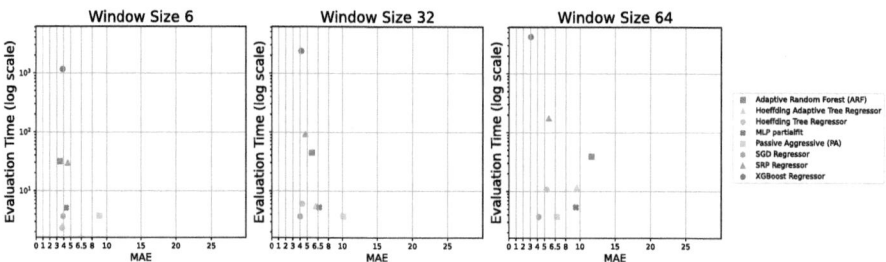

Fig. 4. Prequential evaluation time vs. MAE per model in Experiment II.

4.3 Experiment III

In this experiment, we evaluate Lag-Llama in a similar setting to the previous experiments. The summary results for the window sizes (WS) with the lowest MAE are marked in bold in Table 3.

Lag-Llama is evaluated there for zero-shot and four fine-tuned versions to understand the current state of time-series foundation models for evolving data streams. The original Lag-Llama implementation is primarily designed for forecasting single or multi-step-ahead predictions iteratively rather than for evaluating incoming data streams over time. To address this, we use each model's context length to represent the number of lags for each prediction. A data stream is then simulated over the evaluation set to perform a prequential evaluation for the fine-tuned version of Lag-Llama and compare results to Experiment II.

The experiment involved fine-tuning Lag-Llama models with lags of 32, 64, 128, and 256 (or window sizes) and also testing the same amounts as context lengths (*CL*) in Lag-llama. Simultaneously, RoPE [39], which utilizes rotatory positional embeddings (RoPE) scaling, is assessed. RoPE is evaluated to understand the relative position of lags within the series.

Runtimes of Lag-Llama for fine-tuning range between [1070, 1850] seconds. The mean evaluation time ranged between [89, 107] seconds. All these times have

Table 3. Results of Experiment III. Performance Comparison of MAE, SMAPE, MASE, and R^2 Metrics for Zero-shot and Fine-tuning Approaches on the CPU Dataset. Abbreviations - CL: context length

Model	CL	RoPE	MAE		RMSE		R^2		SMAPE		MASE	
			mean	std	mean	std	mean	std	mean	std	mean	std
Zero shot	32	No	6.252	0.016	14.587	0.088	0.753	0.002	26.692	0.043	2.248	0.014
		Yes	6.249	0.010	14.566	0.015	0.753	0.001	26.661	0.054	2.247	0.015
	64	No	9.819	0.020	19.089	0.027	0.583	0.001	35.981	0.117	3.237	0.015
		Yes	9.355	0.017	18.848	0.021	0.576	0.001	32.656	0.071	3.231	0.017
	128	No	8.847	0.021	15.142	0.041	0.737	0.001	37.356	0.138	2.129	0.013
		Yes	7.112	0.019	14.304	0.037	0.771	0.001	33.109	0.065	1.759	0.010
	256	No	9.651	0.021	14.304	0.044	0.747	0.001	39.567	0.148	1.949	0.011
		Yes	**5.500**	**0.021**	**11.579**	**0.034**	**0.857**	**0.001**	**32.021**	**0.169**	**1.169**	**0.004**
Fine-tuned model on 32 lags	32	No	5.393	0.694	10.651	0.488	0.844	0.020	24.775	1.091	2.106	0.306
		Yes	5.271	0.645	10.703	0.742	0.844	0.025	24.460	0.783	2.037	0.238
	64	No	4.941	0.623	8.482	0.792	0.905	0.020	24.432	1.005	1.558	0.323
		Yes	4.967	0.708	8.439	0.867	0.906	0.022	24.370	1.211	1.463	0.386
	128	No	4.184	0.441	7.204	0.502	0.937	0.010	23.652	0.959	1.128	0.108
		Yes	4.114	0.322	7.274	0.344	0.937	0.006	23.504	0.571	1.095	0.091
	256	No	3.623	0.324	7.074	0.277	0.942	0.005	22.585	0.758	0.904	0.078
		Yes	**3.567**	**0.150**	**7.053**	**0.211**	**0.942**	**0.004**	**22.615**	**0.586**	**0.905**	**0.040**
Finetuned model on 64 lags	32	No	5.184	0.517	10.548	0.466	0.851	0.016	24.852	0.937	1.849	0.142
		Yes	5.383	0.631	10.850	0.733	0.841	0.022	24.566	1.062	2.063	0.304
	64	No	4.838	0.644	8.345	0.548	0.910	0.013	24.469	1.101	1.362	0.235
		Yes	4.778	0.675	8.132	0.726	0.912	0.020	24.722	1.093	1.599	0.289
	128	No	3.817	0.283	6.945	0.342	0.942	0.007	22.968	0.528	1.094	0.111
		Yes	3.881	0.329	7.233	0.647	0.937	0.013	23.021	0.705	1.073	0.104
	256	**No**	**3.514**	**0.161**	**7.158**	**0.211**	**0.940**	**0.004**	**22.460**	**0.639**	**0.896**	**0.048**
		Yes	3.623	0.150	7.316	0.314	0.939	0.005	22.310	0.491	0.922	0.044
Finetuned model on 128 lags	32	No	5.389	0.466	11.125	0.679	0.837	0.020	25.415	0.962	1.671	0.138
		Yes	5.045	0.438	10.922	1.055	0.848	0.028	23.963	0.964	1.921	0.258
	64	No	3.667	0.133	7.034	0.291	0.941	0.005	22.934	0.554	1.027	0.062
		Yes	5.034	0.605	9.136	1.080	0.890	0.028	24.550	0.755	1.658	0.350
	128	No	3.733	0.202	7.063	0.323	0.940	0.006	23.085	0.478	1.045	0.054
		Yes	3.768	0.274	7.145	0.304	0.939	0.006	22.928	0.734	1.088	0.089
	256	No	3.688	0.197	7.562	0.287	0.935	0.005	22.168	0.369	0.882	0.046
		Yes	**3.653**	**0.149**	**7.680**	**0.262**	**0.933**	**0.005**	**22.475**	**0.507**	**0.929**	**0.035**
Finetuned model on 256 lags	32	No	6.927	0.830	13.091	1.349	0.769	0.048	28.311	1.523	1.894	0.180
		Yes	5.278	0.537	12.237	1.565	0.826	0.043	24.532	1.289	1.860	0.249
	64	No	5.678	0.492	12.289	1.339	0.810	0.038	25.260	0.856	1.960	0.240
		Yes	4.586	0.358	9.745	2.001	0.877	0.050	24.048	0.446	1.572	0.270
	128	No	3.881	0.341	7.611	0.434	0.930	0.009	22.948	0.776	1.039	0.102
		Yes	3.821	0.294	7.537	0.771	0.933	0.013	22.851	0.811	1.075	0.094
	256	No	3.740	0.185	7.843	0.356	0.929	0.006	22.685	0.627	0.912	0.041
		Yes	**3.683**	**0.176**	**7.444**	**0.261**	**0.935**	**0.004**	**22.872**	**0.462**	**0.971**	**0.030**

been captured using a GPU (unlike Experiments I and II). Thus, it performs worse when compared to the previous experiments that were computed using the CPU.

From the results observed in Experiment III (see Table 3), it is clear that, in comparison to Experiment II, none of the Lag-Llama tests in our study (neither the zero-shot nor the fine-tuned) were able to outperform ARF or the re-trained XGBoost from Experiment II.

5 Discussion

In this work, choosing the best model across experiments involves a trade-off between performance and computational time. Finding the best model depends on the need for model updates and constraints in the devices needing to predict such workloads. In this study, models are targeted for constrained devices that need low computational inference times. In Experiment I, despite RF showing the best performance, XGBoost obtains very similar results at a lower computational cost. In Experiment II, XGBoost exhibits the highest predictive performance, although it comes with a considerable evaluation time, allowing ARF to take the lead in terms of performance metrics. Nevertheless, ARF still shows a relatively high memory consumption, although this should not be a concern for many edge device setups.

In this work, ensemble models have shown the best overall predictive accuracies and the best tradeoff to computational cost. Online learners in Experiment II have still been able to compete with results from Experiment I but have not been able to overperform them. Models from Experiment II require fewer computational resources compared to deep learning methods in Experiment I or Lag-Llama in Experiment III, which will perform well at the edge when having access to GPU resources. While Lag-Llama is trained on extensive context lengths, it may encounter difficulties in accurately adapting to changes in evolving streams. Furthermore, the algorithms tested in Experiment I consistently outperform Lag-Llama, which largely mirrors the performance of online learners in Experiment II (ARF, HAT, HT) but with larger runtimes. Despite its low pre-training time in Experiment II and the fact that it obtained the best predictive accuracy in our experiments, XGBoost does not scale when being continuously retrained. This is understandable as the algorithm has not been designed for this purpose, and running an adaptive version should be a future line of work. As far as we know, this has not yet been implemented in the software (*River*); hence, this work is out of our scope.

In summary, online learners offer promising results, and a more in-depth study adding extra algorithms and hyperparameters may help find an optimal method. In the meantime, ensembles in Experiment I seem to be the best option for predicting CPU loads in edge devices. A more extensive study using data stream learning benchmarks should be made for this purpose, but it is considered out of scope in this work.

6 Conclusion

This paper has presented an approach to predicting CPU utilization and allowing model selection between state-of-the-art, online ML methods and the time-series foundation model Lag-Llama. The results show promising results for online ML methods and underscore the use of non-linear methods like ensembles or neural networks in case of having access to GPUs to predict CPU load at the edge. The results obtained enforce the relevance of the dataset generated for data stream learning. Our study highlights the effectiveness of online ML methods as a suitable approach for CPU performance estimation. Future research should explore additional applications and extend the proposed evaluation framework to other domains.

Acknowledgments. Our advancement has been made possible by funding from the European Union's HORIZON research and innovation program (Grant No. 101070177).

We are grateful to our colleagues at the EU Horizon project ICOS and CeADAR (Ireland's National Centre for Applied AI) for helping to start and shape this research effort.

Author contributions. S.C., J.S., and A.L.S.-C. collaborated on publishing the data and writing the paper. S.C. handled formal analysis, data curation, and model training for Experiments 2 and 3, while J.S. managed training, evaluation, and data curation for Experiment 1. A.L.S.-C. contributed to conceptualization, methodology, and bibliography review. A.L.S.-C. and R.S.C. supervised the work and reviewed the paper. All authors read and approved the final manuscript.

Disclosure of Interests. The authors have no competing interests to declare that are relevant to the content of this article.

References

1. Tsymbal, A.: The problem of concept drift: definitions and related work. Comput. Sci. Dept. Trinity Coll. Dublin **106**(2), 58 (2004)
2. Suárez-Cetrulo, A.L., Quintana, D., Cervantes, A.: A survey on machine learning for recurring concept drifting data streams. Expert Syst. Appl. **213**, 118934 (2023)
3. Gomes, H.M., et al.: Adaptive random forests for evolving data stream classification. Mach. Learn., 1469–1495 (2017). https://doi.org/10.1007/s10994-017-5642-8
4. Gama, J., Sebastiao, R., Rodrigues, P.P.: On evaluating stream learning algorithms. Mach. Learn. **90**, 317–346 (2013)
5. Zhang, Y., Sun, W., Inoguchi, Y.: Predict task running time in grid environments based on CPU load predictions. Futur. Gener. Comput. Syst. **24**(6), 489–497 (2008)
6. Bey, K.B., Benhammadi, F., Mokhtari, A., Guessoum, Z.: CPU load prediction model for distributed computing. In: Eighth International Symposium on Parallel and Distributed Computing, pp. 39–45. IEEE (2009)
7. Mason, K., Duggan, M., Barrett, E., Duggan, J., Howley, E.: Predicting host CPU utilization in the cloud using evolutionary neural networks. Futur. Gener. Comput. Syst. **86**, 162–173 (2018)
8. Duggan, M., Mason, K., Duggan, J., Howley, E., Barrett, E.: Predicting host CPU utilization in cloud computing using recurrent neural networks. In: 12th International Conference for Internet Technology and Secured Transactions (ICITST), pp. 67–72. IEEE (2017)

9. Janardhanan, D., Barrett, E.: CPU workload forecasting of machines in data centers using LSTM recurrent neural networks and ARIMA models. In: 12th International Conference for Internet Technology and Secured Transactions (ICITST), pp. 55–60. IEEE (2017)
10. Karim, M.E., Maswood, M.M.S., Das, S., Alharbi, A.G.: BHyPreC: a novel Bi-LSTM based hybrid recurrent neural network model to predict the CPU workload of cloud virtual machine. IEEE Access **9**, 131476–131495 (2021)
11. Valarmathi, K., Kanaga Suba Raja, S.: Resource utilization prediction technique in cloud using knowledge based ensemble random forest with LSTM model. Concurr. Eng. **29**(4), 396–404 (2021)
12. Makridakis, S., Spiliotis, E., Assimakopoulos, V.: Statistical and machine learning forecasting methods: concerns and ways forward. PLoS ONE **13**(3), e0194889 (2018)
13. Borovykh, A., Bohte, S., Oosterlee, C.W.: Dilated convolutional neural networks for time series forecasting. J. Comput. Fin. (2018)
14. Patel, E., Kushwaha, D.S.: A hybrid CNN-LSTM model for predicting server load in cloud computing. J. Supercomput. **78**(8), 1–30 (2022)
15. Kiranyaz, S., Avci, O., Abdeljaber, O., Ince, T., Gabbouj, M., Inman, D.J.: 1D convolutional neural networks and applications: a survey. Mech. Syst. Sig. Process. **151**, 107398 (2021)
16. Wang, Z., Yan, W., Oates, T.: Time series classification from scratch with deep neural networks: a strong baseline. In: International Joint Conference on Neural Networks (IJCNN), pp. 1578–1585. IEEE (2017)
17. Hochreiter, S., Schmidhuber, J.: Long short-term memory. Neural Comput. **9**(8), 1735–1780 (1997)
18. Hsu, M.-W., Lessmann, S., Sung, M.-C., Ma, T., Johnson, J.E.: Bridging the divide in financial market forecasting: machine learners vs. financial economists. Expert Syst. Appl. **61**, 215–234 (2016)
19. Rasul, K., et al. Lag-llama: Towards foundation models for probabilistic time series forecasting. arXiv preprint arXiv:2310.08278 (2024)
20. Farahnakian, F., Pahikkala, T., Liljeberg, P., Plosila, J., Hieu, N.T., Tenhunen, H.: Energy-aware VM consolidation in cloud data centers using utilization prediction model. IEEE Trans. Cloud Comput. **7**(2), 524–536 (2016)
21. Ruder, S.: An overview of gradient descent optimization algorithms. arXiv preprint arXiv:1609.04747 (2016)
22. Shaikh, R., Muntean, C.H., Gupta, S.: Prediction of resource utilisation in cloud computing using machine learning. In: CLOSER, pp. 103–114 (2024)
23. Breiman, L.: Classification and Regression Trees. Routledge (2017)
24. Suárez Cetrulo, A.L.: Adaptive algorithms for classification on high-frequency data streams: application to finance (2022)
25. Rahmanian, A.A., Ghobaei-Arani, M., Tofighy, S.: A learning automata-based ensemble resource usage prediction algorithm for cloud computing environment. Futur. Gener. Comput. Syst. **79**, 54–71 (2018)
26. Chen, T., Guestrin, C.: XGBoost: a scalable tree boosting system. In: Proceedings of the 22nd ACM SIGKDD International Conference on Knowledge Discovery and Data Mining, pp. 785–794 (2016)
27. Iqbal, W., Berral, J.L., Erradi, A., Carrera, D., et al.: Adaptive prediction models for data center resources utilization estimation. IEEE Trans. Netw. Serv. Manage. **16**(4), 1681–1693 (2019)

28. Kim, I.K., Wang, W., Qi, Y., Humphrey, M.: Cloudinsight: utilizing a council of experts to predict future cloud application workloads. In: IEEE 11th International Conference on Cloud Computing (CLOUD), pp. 41–48. IEEE (2018)
29. Drucker, H., Burges, C.J., Kaufman, L., Smola, A., Vapnik, V.: Support vector regression machines. Adv. Neural Inf. Process. Syst. **9** (1996)
30. Hu, R., Jiang, J., Liu, G., Wang, L.: CPU load prediction using support vector regression and kalman smoother for cloud. In: IEEE 33rd International Conference on Distributed Computing Systems Workshops, pp. 88–92. IEEE (2013)
31. Crammer, K., Dekel, O., Keshet, J., Shalev-Shwartz, S., Singer, Y., Warmuth, M.K.: Online passive-aggressive algorithms. J. Mach. Learn. Res. **7**(3) (2006)
32. Moghar, A., Hamiche, M.: Stock market prediction using LSTM recurrent neural network. Procedia Comput. Sci. **170**, 1168–1173 (2020)
33. Domingos, P., Hulten, G.: Mining high-speed data streams. In: Proceedings of the Sixth ACM SIGKDD International Conference on Knowledge Discovery and Data Mining, pp. 71–80 (2000)
34. Bifet, A., Gavalda, R.: Adaptive learning from evolving data streams. In: Advances in Intelligent Data Analysis VIII: 8th International Symposium on Intelligent Data Analysis, IDA 2009, Lyon, France, August 31-September 2, 2009. Proceedings 8, Springer, pp. 249–260 (2009)
35. Bifet, A., Holmes, G., Pfahringer, B.: Leveraging bagging for evolving data streams. In: Machine Learning and Knowledge Discovery in Databases: European Conference, ECML PKDD 2010, Barcelona, Spain, September 20-24, 2010, Proceedings, Part I 21, Springer, pp. 135–150 (2010)
36. Chen, S.-T., Lin, H.-T., Lu, C.-J.: An online boosting algorithm with theoretical justifications. arXiv preprint arXiv:1206.6422 (2012)
37. Gomes, H.M., Read, J., Bifet, A.: Streaming random patches for evolving data stream classification. In: IEEE International Conference on Data Mining (ICDM), pp. 240–249. IEEE (2019)
38. Botchkarev, A.: Performance metrics (error measures) in machine learning regression, forecasting and prognostics: properties and typology. arXiv preprint arXiv:1809.03006 (2018)
39. Su, J., Ahmed, M., Lu, Y., Pan, S., Bo, W., Liu, Y.: RoFormer: enhanced transformer with rotary position embedding. Neurocomputing **568**, 127063 (2024)

RoCoNA: A Robust Continual Learning Framework for Alignment of Dynamic Networks Under Distribution Shift and Domain Differences

Shruti Saxena[✉] and Joydeep Chandra

Indian Institute of Technology Patna, Patna, India
{shruti_2021cs31,joydeep}@iitp.ac.in

Abstract. Network alignment, which maps the same entities across multiple networks, has gained tremendous interest in recent years. However, most existing alignment methods propose to align static graphs that are merely a single snapshot of real-world networks. These methods fail to model the inherent dynamics of entire networks where the nodes, links, and attributes are bound to change over time, making dynamic network alignment challenging. Moreover, modeling the interaction between two different dynamic graphs comes with additional challenges: (1) catastrophic forgetting while learning the evolution of individual networks, (2) distributional shift on the same dynamic network, and (3) domain differences between the two networks. Hence, to overcome these challenges, we propose RoCoNA, an end-to-end reservoir sampling-based continual learning approach built over streaming Graph Neural Networks (GNNs) that uses a novel shift-induced regularizer to handle distribution drift and domain differences in evolving networks. We empirically show that our method outperforms the existing state-of-the-art static and dynamic alignment methods. We perform case studies on networks with high distributional shifts to strongly validate our claims. Code is available at https://github.com/shruti400/RoCoNA/tree/main

Keywords: Network alignment · Continual learning · Distribution shift

1 Introduction

In the era of Big-data, graphs are one of the most compelling ways of representing and analyzing complex real-world interactions across several domains, such as social networks, co-authorship, and protein interaction networks. More often, the interacting entities in each network (like the persons, authors, and proteins) appear in multiple networks and exhibit different interactional behavior [13]. For example, a user's interaction style across online social networks, like Facebook, Twitter, or LinkedIn, may vary. The participation of a protein in several genetic pathways is also analogous. Network alignment, or mapping the same entities across several networks, can provide these entities a broader feature, benefiting various intriguing applications, including cross-species gene prioritization, fraud detection, and recommendation engines [2].

Comprehensive network alignment research is built upon static networks and focuses on preserving the structural and attribute consistency of the partially aligned networks [3,12,16,21,24,25]. Despite their remarkable results, these methods fail to account for the inherent dynamics of the continuously evolving real-world networks. For example, edges are added or removed in social networks, and the attributes of nodes may also change over time. These dynamics give rise to some new patterns while retaining some existing ones. Markedly, network dynamics can help in network alignment by capturing the trends in data [1].

Motivated to capture the network dynamicity for better results in the real-world scenario, we study the problem of aligning entities in two dynamic social networks in this paper. We simultaneously deal with significant challenges that remain in modeling the interaction between two different dynamic networks:

1. *How to learn the evolving behavior of dynamic networks?* Nodes or edges are added or removed in dynamic networks, and the attributes of nodes may also change over time. These dynamics give rise to new patterns while retaining existing ones, and capturing these network dynamics remains challenging.
2. *How to consolidate the natural phenomenon of distributional shift on dynamic networks?* The network structure and node feature distribution of large networks like social networks and citation networks follow a strong time-dependent correlation [8]. For instance, the distributional characteristics of a citation network would go through significant change as new research fields grow, and authors may generate distinct types of social circles in different years.
3. *How to align networks that come from different domains?* The source and target networks can be of different domains with different evolution rates and patterns, and matching their distributional spaces for generalizing the dynamic behavior of both networks remains challenging.

Existing dynamic alignment methods [18,19] typically focus on the first and third challenges, capturing the network dynamics via an LSTM Autoencoder framework and matching the domains by constructing a common subspace guided by the pre-known correspondences, a.k.a. anchors. However, these methods overlook the significant shift in the data distributions resulting from the temporal augmentations in an evolving network. To the best of our knowledge, we are the first to handle the distribution shifts in the face of dynamic network alignment. Moreover, the existing LSTM-based methods keep learning new patterns, overwriting, and abruptly forgetting the existing patterns, making them prone to the problem of *catastrophic forgetting.*

Given the above challenges and the need to collectively address them, we propose RoCoNA, an end-to-end continual learning-based Graph Neural Network (GNN) model for dynamic network alignment. We use a memory reservoir-based strategy to keep track of the evolution behavior coupled with a novel distribution divergence regularizer to handle distribution drift and domain differences. Since memory capacity is limited, we propose both, a novel PageRank-induced sampling strategy for identifying representative interactions to store in the memory reservoir, and an online memory updation technique that eliminates the least representative interactions from memory. The distribution divergence regularization is a statistical moment-matching approach involving two strategies: hard distribution matching for addressing the network's distribution shift and soft distribution matching for addressing the domain differences. The end-to-end framework addresses the problem of catastrophic forgetting by ensuring that the model trained over the current timestep still applies for earlier timesteps. It also effectively captures new patterns emerging due to the distributional shift of evolving networks. We empirically validate the effectiveness of RoCoNA on several real-world datasets and compare it to several state-of-the-art static and dynamic network alignment methods.

2 Related Works

Network alignment is the task of identifying the same entities across multiple networks on their common users *a.k.a.* anchors. The earliest works include spectral methods based on a matrix decomposition formulation [10,17,27]. More recently, deep learning-based methods propose joint learning of the two networks optimized over a loss function. Methods like [3,6,7] impose strong topological constraints that strictly impose the anchors to be structurally and semantically the same, whereas some methods use noise and adversarial robustness, like contrastive learning [4,16,23] and adversarial learning-based approaches [7,25,28] to better capture the network variances. However, all these approaches consider networks to be static and fail to exploit their underlying dynamics. Existing studies on dynamic network alignment primarily focus on biological domains where they find similar conserved regions between two networks for function prediction [1,22]. Hence, they differ from our goal of finding entity mappings. The problem of aligning dynamic social networks has not yet been thoroughly explored, with only a few proposed works. DNA [18] is the first framework that captures each node's dynamic and global consistency for aligning users across dynamic networks. It proposes an LSTM Autoencoder framework for capturing the intra-network dynamics while constructing a common subspace for inter-network alignment. DGA [19] extends it by replacing the Random walk with restart (RWR) encoding module with a graph attention method, reducing DNA's convergence time and parameter size. HDyNA [5] distinguishes itself by proposing a heuristic algorithm that locally updates the weight of only newly arrived nodes during the network evolution.

3 Problem Formulation

Definition 1 (Network as Graph). *We formally define a network $G = (V, \hat{A}, X)$ as an undirected, unweighted graph with a node set $V = \{v_1, v_2, ..v_n\}$, adjacency matrix $\hat{A} \in \{0,1\}^{n \times n}$, and an attribute matrix $X \in \mathbb{R}^{n \times m}$ where $x_i \in \mathbb{R}^m$ represents the feature vector of the node v_i.*

Definition 2 (Dynamic Network). *A dynamic network is a sequence of static graph snapshots, $\mathcal{G} = \{G^0, G^1, ..., G^T\}$, with T timestamps. At each time step $t = \{1, 2, .., T\}$, $G^t = (V^t, \hat{A}^t, X^t)$.*

*Problem 1 (**Dynamic Network Alignment**).*
Given: *(1) two dynamic networks $\mathcal{G}_s = \{G_s^0, G_s^1, .., G_s^T\}$ and $\mathcal{G}_\tau = \{G_\tau^0, G_\tau^1, .., G_\tau^T\}$, and (2) a set of anchor node pairs A'.*
Output: *An alignment matrix $A_t^\star$, $\forall t \in T$ where $A_t^\star(u, v)$ represents the similarity between nodes $u \in V_s^t$ and $v \in V_\tau^t$.*

Here, A' is of fixed size, but the size of A^t varies over time.

3.1 Preliminaries: Graph Neural Networks

A typical L-layer GNN architecture iteratively aggregates information from its neighborhood using a learnable aggregator function F^θ, where θ are the learnable weight parameters. Successive layers are stacked together to generate node representations as:

$$Z^{(l)} = F^\theta(\tilde{A}Z^{(l-1)}; \theta) \quad (1)$$

where $Z^{(l)} \in [a, b]^n$, $Z^{(0)} = X$ and $Z^{(l)}(i) = \boldsymbol{h}_i^{(l)}$ is the representation of a node i at the lth layer. We represent the final node representations $Z^{(L)}$ as Z throughout the work for the sake of simplicity. $\tilde{A} = D^{-\frac{1}{2}}(\hat{A} + I)D^{-\frac{1}{2}}$ and D is the degree matrix of $(\hat{A} + I)$. Moreover, graph pooling is adopted to summarize node embeddings for representing the entire graph as:

$$\boldsymbol{z} = \frac{1}{n}\sum_{j=1}^{n} Z_{:j}^T \quad (2)$$

Streaming GNNs. are extensions of traditional GNNs in a dynamic environment. Given a dynamic network $\mathcal{G}$, the goal is to learn learn $(\theta^1, \theta^2, \ldots, \theta^T)$ where θ^t is the parameters of the GNN F at time t. We represent the final latent node representations at any time t as $Z^t = \{\boldsymbol{h}_0^t, \boldsymbol{h}_1^t, .., \boldsymbol{h}_n^t\}$ and the graph representation as $\boldsymbol{z}^t$.

4 Methodology

We build RoCoNA over a continual learning framework for generating node representations of the source and target networks by minimizing the following loss functions: (a) a continual loss function for adapting to the networks' time-induced structural changes and (b) an anchor alignment loss function for preserving the anchor node embedding similarities in the face of distribution shifts in the individual networks. The following sections discuss the generation of continual node representations, aligned representation, and optimization and alignment computation in detail.

4.1 Continual Representations

The dynamic network environment encounters continuous changes in the network patterns, and we aim to learn representations modeled by streaming GNNs that can capture these changes. We propose streaming GNNs via continual learning that learns (1) new patterns that may occur at each timestamped snapshot and (2) historical patterns from previous snapshots to alleviate catastrophic forgetting. We next explain the proposed learning strategy in detail.

Detecting New Patterns: The new node set, ΔV^t, of consecutive snapshots can be intuitively regarded as new patterns. However, even nodes other than ΔV^t can depict new patterns as some changes in the new nodes can affect the interactional behaviors of the existing nodes, and these nodes also need to be retrained. Hence it is crucial to mine the set of nodes that show new patterns during network evolution.

We define new patterns based on the differences in the PageRanks of nodes at each consequent snapshot. A high difference in node ranks implies the occurrence of unusual interactions during $[t-1, t]$ and hence reflects their importance. The underlying intuition is that the dynamic PageRank of ordinary nodes whose neighborhoods are not significantly affected does not change much, and it is unnecessary to recompute their representation [15]. Hence the set of affected nodes between two consecutive snapshots is given by

$$\mathcal{I} = \Delta V^t + \{\|\mathbf{r}_u^t - \mathbf{r}_u^{t-1}\| > \delta\}_{\forall u \in V^t \cap V^{t-1}}, \qquad (3)$$

where $\mathbf{r}_u^t$ is the PageRank score of a node u at time t. δ is a hyper-parameter that controls the number of nodes to be treated as new patterns. We utilize [15] for efficiently calculating the dynamic PageRank scores.

Preserving Existing Patterns: We preserve the existing patterns from the previously timestamped snapshots by following the reservoir sampling technique. We take a fixed-size memory, say $\mathcal{M}$, and sample only the most representative interactions from each snapshot. At any time t, the idea is to make the model learn additionally from the consolidated historical information preserved in the reservoir memory.

Therefore, we propose an importance-based sampling strategy based on the PageRank node centrality scores that quantify the relative influence of graph nodes. We scale the constant probability of sampling a random node, p_u^t, in the reservoir memory based on its importance factor. The underlying intuition is that the common samples have limited importance with respect to the model convergence and representing patterns, and hence are more likely to increase the redundant information in the reservoir memory. The scaled probability is calculated as follows:

$$p_{u \leftarrow \mathcal{M}}^t = 1 - \min\left(p_u^t \frac{r_{\max}^t - r_u^t}{c_{\max} - r_{mean}^t}, p_{\mathcal{M}}^t\right), \qquad (4)$$

where r_{max}^t and r_{mean}^t are the maximum and average PageRank scores over all nodes, and $p_\mathcal{M}^t$ is a hyperparameter that controls the amount of memory $\mathcal{M}$ updated by the samples at time t. Since the memory is of fixed size, we propose an online memory-updating strategy that discards the nodes with the most stable PageRank scores across all the previous snapshots whenever it gets full. So, at time t, we calculate the probability of a node u being removed from memory by:

$$p_{\mathcal{M}\leftarrow u}^t = \frac{t - t_u^\mathcal{M}}{\frac{1}{t}\sum_{t'\in\{t,t-1,\ldots 1\}} |r_u^{t'} - r_u^{t'-1}|}, \tag{5}$$

where $t_u^\mathcal{M}$ is the time at which node u was sampled into the memory.

At each timestamp, the encoder model is trained over the newly detected patterns and the sampled patterns in the memory. Our aim is to learn a GNN parameterized by θ on G^t while also maintaining low errors in the previously timestamped snapshots. To achieve this goal, we follow the Bayesian approach as used in EWC [9] and devise the continual loss function as:

$$\mathcal{L}_t^{CL} = \mathcal{L}_t^{cons} + \sum_x \frac{\lambda}{2} F_x^t \left(\theta_x^t - \theta_x^{(t-1)*}\right)^2 \tag{6}$$

The regularization term solves the overfitting problem caused by replaying the small number of nodes in the fix-sized reservoir memory. F_x^t is the Fisher Information matrix of the xth parameter of θ and is computed from the first order derivatives. $\theta_x^{(t-1)*}$ represents the optimal set of parameters of the previous snapshot that lead to the least errors, and λ regularizes the importance of information transfer between consecutive snapshots. $\mathcal{L}_t^{cons}$ is the consistency loss for G^t only that ensures that the learned node embeddings are consistent with the network topology of G^t and is given by:

$$\mathcal{L}_t^{cons} = \sum_{i\in\mathcal{M}\cup I} l\left(\theta^t; i\right) = \sum_{i\in\mathcal{M}\cup I} \left\|\tilde{A}^t(i) - \sigma\left(\boldsymbol{h}_i^t(\boldsymbol{h}_i^t)^T\right)\right\|_F, \tag{7}$$

where $\|\cdot\|_F$ denotes the Frobenius norm. The loss is computed over the embeddings at all layers to ensure node neighborhood consistency at different orders.

4.2 Aligned Representations

We leverage the pre-known anchor mappings A' to align the latent representations of the two networks. We increase the similarity between the anchor nodes of each timestamped snapshot pair, (G_s^t, G_τ^t), using a negative sampling strategy. We define the anchor alignment loss function as:

$$\mathcal{L}_t^{anchor} = \frac{1}{|A_t|} \sum_{(i,j)\in A_t} \left(s\left(i,j\right) + \frac{1}{|U_t|} \sum_{(m,n)\in U_t} (1 - s\left(m,n\right))\right) \tag{8}$$

where U_t is the set of negative anchor links sampled from the unmapped links across the two networks at time t and $s(i,j)$ gives a measure of similarity between nodes i and j as:

$$s(i,j) = \sigma\left((\mathbf{h}_i)^T \cdot \mathbf{h}_j\right) \qquad (9)$$

However, we require more than relying on the anchor alignment loss while dealing with dynamic networks of different domains, evolution rates, and data distributions, where the alignment efficacy directly depends on how effectively we handle these differences. Hence, we utilize a statistical moment-matching approach involving two strategies- hard matching and soft matching, for distribution matching with different emphases to different timestamped snapshots. While hard matching is for matching the distributions of the closely time-stamped snapshots, soft matching effectively accounts for the networks' entire evolving behavior.

Hard Distribution Matching: We use the central moment discrepancy (CMD) metric [26] to directly maximize the similarity between the distribution of snapshots G_s^t, G_τ^t, G_s^{t-1}, and G_t^{t-1} at time t. CMD matches higher-order moments such as variance, skewness, and kurtosis in contrast to the widely used Kullback-Leibler (KL) divergence [26], which only matches the mean. We formulate the hard regularisation term as:

$$\mathcal{L}_t^{hard} = d_{cmd}(Z_s^t, Z_\tau^t) + \sum_{g \in \{s,\tau\}} d_{cmd}(Z_g^t, Z_g^{t-1}) \qquad (10)$$

where $d_{cmd} : [a,b]^n \times [a,b]^n \rightarrow \mathbb{R}^+$ is a distributional shift metric given as:

$$d_{\text{cmd}}\left(Z^t, Z^{t'}\right) = \frac{1}{b-a}\left\|\mathbf{E}(Z^t) - \mathbf{E}(Z^{t'})\right\| + \sum_{k=2}^{\infty}\frac{1}{|b-a|^k}\left\|c_k(Z^t) - c_k(Z^{t'})\right\| \qquad (11)$$

where $\mathbf{E}(Z^t) = \frac{1}{|V^t|}\sum_{i \in V^t} h_i$ and $c_k(Z^t) = \mathbf{E}(Z^t - \mathbf{E}(Z^t))^k$ is the k-th order moment. We use moments up to the 5th order.

Soft Distribution Matching: We propose to minimize the distributional divergence of G^t with the previously timestamped snapshots $\{G^{t-1}, G^{t-2}, .., G^0\}$ by softly learning the consistency between their distributions. The distribution similarity of any $G^{t'}$, $t' < t$, with G^t can be calculated by:

$$p_{g \in \{s,\tau\}}^{t'} = \frac{e^{\cos\left(z_g^t, z_g^{t'}\right)/T}}{\sum_{\hat{t} \in \{t', t'-1...,0\}} e^{\cos\left(z_g^t, z_g^{\hat{t}}\right)/T}} \qquad (12)$$

where z is the graph representation (Eq. 2), $\cos(\cdot,\cdot)$ is the cosine similarity of vectors and T is a temperature parameter. At any time t, we can represent the distribution of all $G^{t'}, t' < t$ using Eq. 12 as

$$\mathbf{P}_{g \in \{s,\tau\}} = [p_g^{t-1}, p_g^{t-2} \cdots, p_g^0] \qquad (13)$$

Here P_s denotes the distribution of the source network snapshots and P_τ of the target network. Now to keep the similarity structure of the dynamic source and target network consistent, we formulate the soft regularisation term as follows:

$$L_t^{soft} = d_{KL}(P_s \| P_\tau) + d_{KL}(P_\tau \| P_s) \tag{14}$$

We here use d_{KL}, the KL divergence, instead of the CMD metric for matching distributions in a soft manner.

Finally, by combining the above two regularization terms, we obtain the goal of distribution alignment as:

$$L_t^{align} = L_t^{anchor} + \alpha L_t^{hard} + \beta L_t^{soft} \tag{15}$$

where α and β are hyper-parameters for weighting the regularization terms.

4.3 Optimization and Alignment

We combine the two loss functions for continual and aligned representations to obtain the objective function at each time-step t as:

$$\mathcal{L} = \mathcal{L}_s^{CL} + \mathcal{L}_\tau^{CL} + \mathcal{L}^{align} \tag{16}$$

We optimize the above objective jointly over both networks in a unified framework using the stochastic gradient descent optimizer. After the model converges, we derive the alignment matrix A_t^* with the generated embeddings as:

$$A_t^* = \sigma\left((Z_s^t)^\top \cdot Z_\tau^t\right) \tag{17}$$

The $(u,v)^{th}$ entry in A_t^* signifies the similarity score between the corresponding nodes. Hence the highest value corresponding to a node u in the row A_t^* most likely represents the same node in the other network.

5 Experimental Details

Datasets: We evaluate the effectiveness of RoCoNA on four real datasets from different fields.

- Social Networks: We use Twitter and Foursquare, two famous social networks connecting users online [12]. We obtain a sequence of snapshots for both networks based on the friending time information of users [19].
- Academic Networks: We take two pairs of academic networks, DBLP-AMiner, and MAG-AMiner. Each is a co-authorship network where author nodes are connected based on their collaborations [20]. We obtain a sequence of snapshots via the first publication time of the collaboration among authors [19].

Table 1. Comparison of network alignment methods on real-world datasets

Datasets	Metric	Static Methods						Dynamic methods				Proposed method
		IONE	CENALP	GAlign	BRIGHT	HackGAN	HCNA	HDyNA	CTSA	DNA	DGA	RoCoNA
TF	Acc@1	11.26	12.32	13.45	15.06	13.01	17.14	16.77	14.98	31.04	31.17	**37.10**
	Acc@10	19.88	20.10	22.67	23.88	21.98	24.62	21.76	22.86	36.64	38.79	**44.65**
	MAP	14.34	14.64	17.11	19.98	15.00	19.12	19.00	18.89	32.37	34.82	**40.30**
TF+	Acc@1	10.06	10.86	12.35	13.45	11.10	15.34	14.39	16.42	30.12	33.58	**36.70**
	Acc@10	18.58	20.14	22.45	22.65	20.76	22.54	20.65	24.67	36.39	39.58	**42.76**
	MAP	15.03	14.73	14.32	15.35	17.92	18.89	17.03	19.41	31.85	34.93	**38.85**
DA	Acc@1	11.28	10.97	11.34	12.04	10.67	14.95	18.76	23.44	32.63	38.88	**40.04**
	Acc@10	20.65	19.84	19.89	20.06	18.96	22.48	22.38	31.69	38.73	42.27	**46.83**
	MAP	12.86	10.47	13.39	14.11	12.03	17.25	19.63	25.48	33.13	39.05	**44.55**
MA	Acc@1	10.96	11.58	13.48	13.01	12.10	15.79	17.84	20.82	32.21	37.17	**41.28**
	Acc@10	19.76	21.44	23.58	22.88	21.47	25.67	25.45	27.57	39.65	43.24	**47.89**
	MAP	12.69	13.10	14.75	14.22	14.15	18.16	18.01	23.36	34.28	39.75	**43.34**

Baselines: We compare RoCoNA with the existing dynamic social network alignment methods, including DNA [18], DGA [19], and HDyNA [5]. We also extend the formulation of CTSA [11], a GNN-based alignment method that originally aligns snapshots of the same network, to align snapshots of different dynamic networks. Moreover, to demonstrate the significance of dynamics in aligning networks, we additionally compare RoCoNA with some recent state-of-the-art static alignment methods, including IONE [12], CENALP [3], GAlign [21], BRIGHT [24], HackGAN [25] and HCNA [16].

Evaluation Metrics: Similar to several high-quality recent works [21,24,25], we compare the approaches using $Acc@q$, which indicates if a node's true anchor match is present in a list of top-q potential anchors. It is given as

$$Acc@q = \frac{\sum_{u_s^* \in V_s} \mathbb{1}_{P^*[u_s^*, u_t^*] \in R(u_s^*)}}{\#\{ \text{ ground truth anchor links } \}} \quad (18)$$

where $(u_s^*, u_t^*) \in A$ and $R(u_s)$ is a list of highest q values in the row $P^*(u_s)$. Apart from this, we also observe the *Mean Average Precision (MAP)* scores [21,25] for ranking perspective.

Implementation Details: We tune the hyper-parameters using the grid search algorithm implemented with Hyperopt. We use a GCN encoder with two hidden layers, $L = 2$, each layer of size 64. We take $\gamma = 0.005$, the size of memory $\mathcal{M}$ as 300 and search $p_{\mathcal{M}}^t$ in the range $\{0.1, 1\}$. The negative sampling size for each anchor in the training set is $K = 10$, and the temperature parameter is $T = 0.5$. We set $\lambda = (40, 30, 20, 20)$, $\alpha = (2, 3, 4, 4)$ and $\beta = (3, 3, 5, 6)$ for TF, TF+, DA and MA datasets respectively. We set the number of training epochs to 200, the embedding dimension to 128, and the initial learning rate to 0.005 for all datasets. For calculating the PageRank using [15], we set the random jump probability $\mu = 0.85$. We use the default values of hyper-parameters for the baselines except for the embedding dimensions for a fair comparison.

6 Results and Analysis

6.1 Comparison with Baselines

We compare RoCoNA to static and dynamic baselines and report the results for $Acc@1$, $Acc@10$, and MAP in Table 1. We leverage five snapshots separated by 80 days, perform alignment at each time interval and then report the averaged result. For comparison with static methods, we follow [19] to generate a ϵ-overlap dataset by deleting users randomly to assess the impact of overlap on entity alignment. We align datasets with 25%, 30%, 35%, 40%, and 45% overlap and report their average. Figure 1 shows the performance of the best-performing methods using 30% training data at each interval (or at each ϵ-overlap for static methods). RoCoNA performs best, improving the alignment accuracy, $Acc@1$ by a minimum margin of 19.08%, 9.28%, 3%, and 11% on the TF, TF+, DA, and MA datasets, respectively. We also observe a minimum improvement of 15.12%, 8.06%, 10.8%, and 10.7% in Acc@10 and 15.7%, 11.2%, 14.1%, and 9% in MAP scores. The results validate the effectiveness of continual learning and distribution matching for superior network alignment.

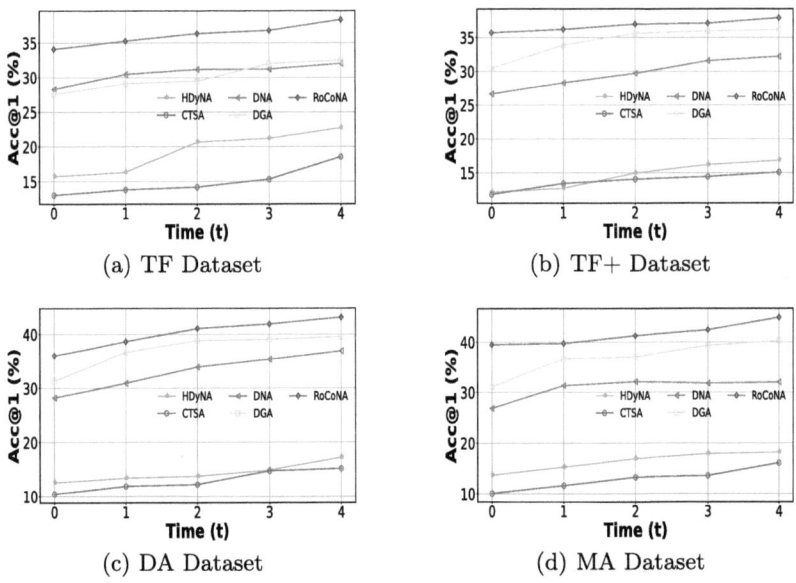

Fig. 1. Acc@1 of the best-performing baselines at each time step.

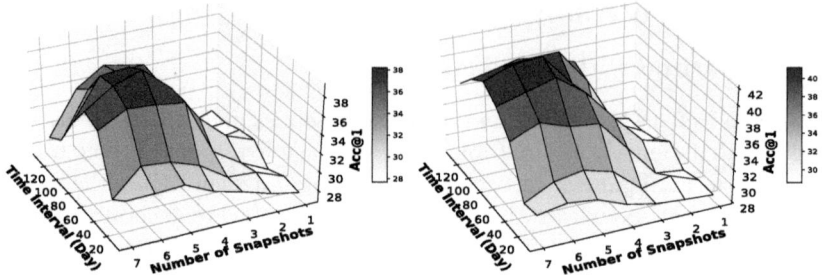

Fig. 2. Snapshotting settings for DA dataset using DGA(left) and RoCoNA(left).

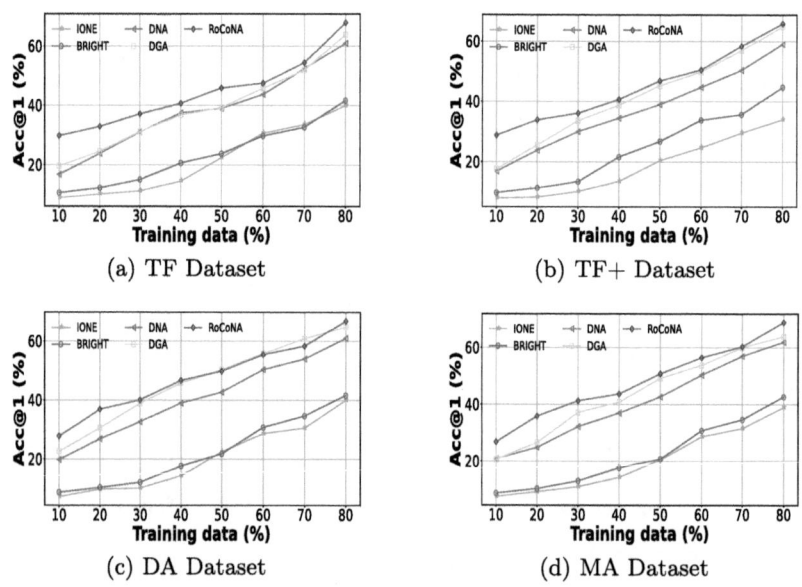

Fig. 3. Alignment results of the supervised baselines under different training rates

6.2 Analysis on the Snapshot Settings

We study the performance of RoCoNA by varying the frequency and number of snapshots and compare it with the best-performing baseline DGA. We report results for the DA dataset in Fig. 2 and observe consistent results for other datasets. We observe that the $Acc@1$ score increases with the number of snapshots and time interval, reaches a maximum for around 5 snapshots with a time interval of 90 days, and then gradually goes down. It shows that only an adequate amount of information is needed to capture the network dynamicity, historical information does not add to its performance, and larger time intervals introduce distribution shifts. DGA, compared to RoCoNA, experiences a drastic performance drop when the time interval goes beyond 85 days, implying that RoCoNA can better handle distribution shifts over time.

6.3 Analysis on the Supervised Information

We observe the performance of RoCoNA and the baselines by varying the amount of supervised anchor information utilized for training. We report the results for all the datasets in Fig. 3. We see that the Acc@1 score of all the methods increases with the increase in training data. RoCoNA outperforms all the methods even at very low training rates, demonstrating that it is effective in capturing network patterns and evolution behavior in the presence of some pre-known correspondences.

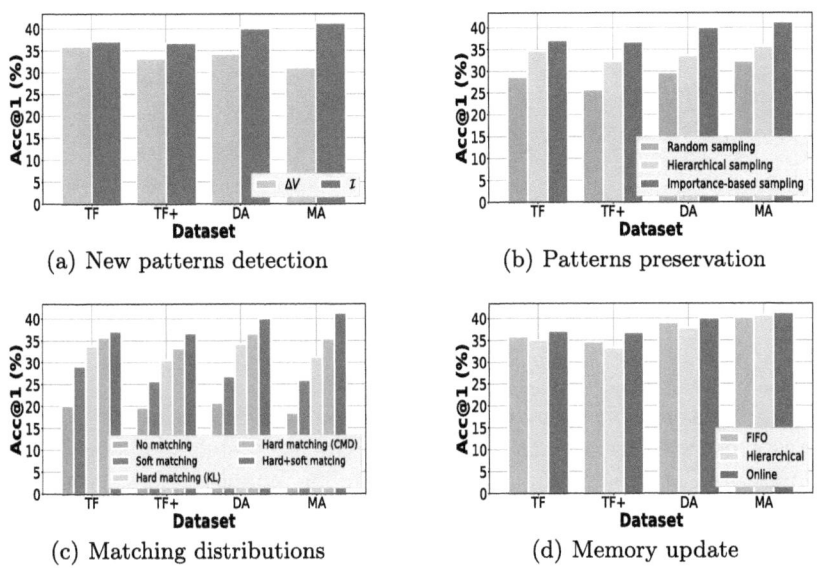

(a) New patterns detection

(b) Patterns preservation

(c) Matching distributions

(d) Memory update

Fig. 4. Analysis of Model design

6.4 Analysis of Model Design

We analyze each proposed component accounting for the effectiveness of RoCoNA in Fig. 4.

New Patterns Detection. We compare the proposed PageRank difference-based technique, which treats $\mathcal{I}$ as in Eq. 3 as new patterns, with the straightforward strategy, which treats ΔV as new patterns. Our proposed method outperforms by $3.13 - 32.74\%$ across all datasets, showing the importance of mining nodes other than ΔV undergoing structural changes during network evolution.

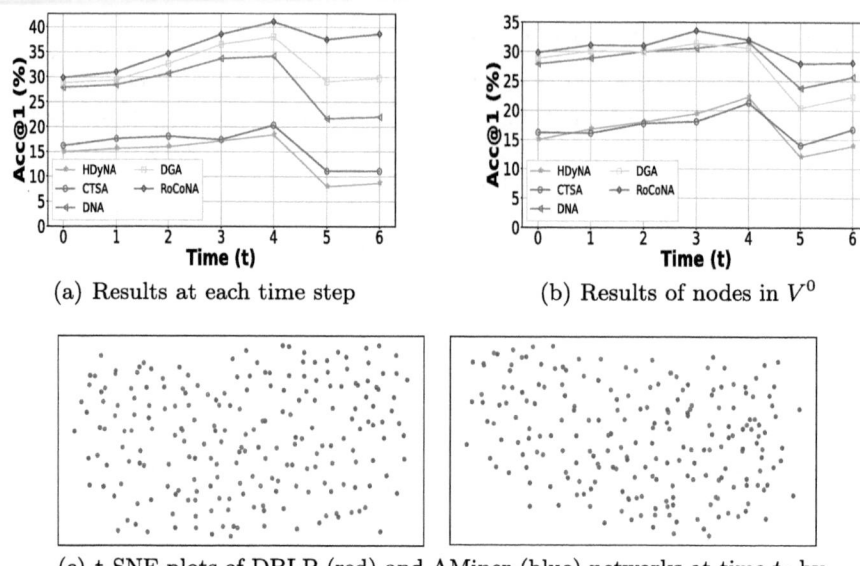

(a) Results at each time step

(b) Results of nodes in V^0

(c) t-SNE plots of DBLP (red) and AMiner (blue) networks at time t_5 by DGA (left) and RoCoNA (right), respectively.

Fig. 5. Case Study

Existing Patterns Preservation. We compare different sampling strategies for preserving existing patterns in memory. Our suggested sampling technique is importance-based sampling. Random sampling samples nodes with uniform probabilities and rank-based sampling samples the top PageRanked nodes. Our importance-based sampling yields better results, proving that it can effectively consolidate historical information compared to other methods.

Matching Distributions. We analyze the effectiveness of our proposed distribution matching strategy. We compare it with no matching, which does not perform any matching, only soft matching (Eq. 14), only hard matching(CMD) (Eq. 10), and only hard matching(KL) that uses KL-divergence instead of the CMD metric. From the results, we see that matching distributions are necessary for dealing with dynamic networks. While matching higher-order moments using the CMD metric in place of KL divergence enhances the alignment accuracy, jointly performing hard and soft matching yields the best result.

Memory Update. We compare the proposed online memory updation strategy in Eq. 5 with the simple 'First In First Out' (FIFO), which discards the earliest sampled nodes when the memory gets full, and the hierarchical strategy, which discards the lowest PageRanked nodes. Results show that online memory updation outperforms all other methods across all datasets.

6.5 Case Study

We conduct experiments on the DA dataset to prove the effectiveness of our model in dealing with catastrophic forgetting and distribution shifts. We take the first five snapshots ($t_0 - t_4$) with a time interval of 90 days, and then at time t_5, we consider a more significant gap of 200 days. Such a split naturally introduces a distribution shift since several latent influential factors, like the popularity of research topics and author collaborations for data generation, would change over time. We then perform case studies to answer the following questions:

(a) *How well is the distribution shift handled?* Figure 5(a) shows the performance of all the dynamic alignment methods in terms of $Acc@1$. We notice that the $Acc@1$ scores of all methods drop significantly at time t_5 as the distribution of the network structure changes. However, the performance of RocoNA remains comparably stable. Figure 5(c) shows the t-SNE visualization of the representations of the anchor nodes (100 anchors randomly extracted) for RoCoNA and the best-performing baseline DGA. We clearly see that as the network patterns change at time t_5, the distribution discrepancy between the source and target networks significantly increases for DGA. At the same time, RoCoNA still achieves a strong correspondence between the two aligned data distributions, making it robust to distribution shifts over time.

(b) *How well is catastrophic forgetting prevented?* Figure 5(b) shows the alignment performance of nodes at time t_0 using the updated GNN model as time evolves. We observe that at the $t_0 - t_2$ time steps, the alignment accuracy remains stable since the distribution of patterns is roughly unchanged. However, as the time interval increases, the GNN gets trained on new nodes and patterns, making the current model no longer applicable to the nodes at t_0. As a result of catastrophic forgetting, the alignment accuracy of all methods rapidly declines. However, RoCoNA shows stable results demonstrating its better generalization capabilities with effective historical knowledge consolidation.

6.6 Computational and Convergence Analysis

We study and compare the execution time and scalability of the network alignment methods in Fig. 6. We follow [14] to use a network-generative model to increase the network size gradually. We observe that RoCoNA converges in a relatively shorter time. HDyNA takes the least time, but its alignment accuracy is not comparable to RoCoNA. We also study the convergence of RoCoNA to total loss ($\mathcal{L}$) on the DA dataset and compare it with the L2 regularised variant, i.e., $\lambda = 0$. From Fig. 7, we see that our regularisation stabilizes better after a fixed number of iterations, indicating the model is convergent towards a similar alignment matrix, preserving historical knowledge constrained by λ.

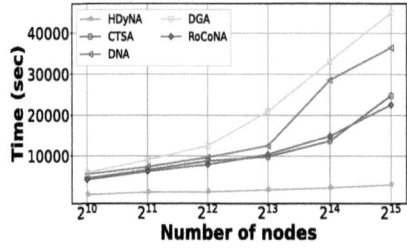 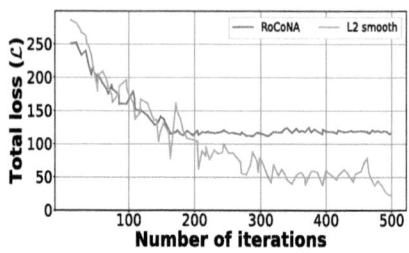

Fig. 6. Computation analysis   **Fig. 7.** Convergence analysis

7 Conclusion

In this paper, we study the dynamic network alignment problem. We focus on addressing the challenges associated when dealing with real dynamic networks- (1) catastrophic forgetting, (2) distribution shift, and (3) domain differences. We bridge them by proposing a continual learning framework, RoCoNA, built over streaming GNNs embodying regularizers to capture networks' evolving behavior. Extensive empirical evaluations show that RoCoNA effectively handles time-induced distribution shifts and exhibits better generalization capabilities by addressing the catastrophic forgetting problem. RoCoNA consistently outperforms the state-of-the-art methods by an up to 19% improvement in Acc@1 across several real datasets. Although we evaluate RoCoNA on limited datasets, its applicability can be extended to other real-world networks, including transactions and biological networks.

References

1. Aparício, D., Ribeiro, P., Milenković, T., Silva, F.: Temporal network alignment via got-wave. Bioinformatics **35**(18), 3527–3529 (2019)
2. Bayati, M., Gleich, D.F., Saberi, A., Wang, Y.: Message-passing algorithms for sparse network alignment. ACM TKDD **7**(1), 1–31 (2013)
3. Du, X., Yan, J., Zha, H.: Joint link prediction and network alignment via cross-graph embedding. In: IJCAI, pp. 2251–2257 (2019)
4. Gao, S., Zhang, Z., Su, S.: DAWN: domain generalization based network alignment. IEEE Trans. Big Data (2022)
5. He, J., Liu, L., Yan, Z., Wang, Z., Xiao, M., Zhang, Y.: User alignment across dynamic social networks based on heuristic algorithm. In: 2021 7th International Conference on Systems and Informatics (ICSAI), pp. 1–7. IEEE (2021)
6. Heimann, M., Shen, H., Safavi, T., Koutra, D.: REGAL: representation learning-based graph alignment. In: Proceedings of the 27th ACM International Conference on Information and Knowledge Management, pp. 117–126 (2018)
7. Hong, H., Li, X., Pan, Y., Tsang, I.: Domain-adversarial network alignment. IEEE Trans. Knowl. Data Eng. (2020)
8. Hu, W., et al.: Open graph benchmark: datasets for machine learning on graphs. Adv. Neural. Inf. Process. Syst. **33**, 22118–22133 (2020)

9. Kirkpatrick, J., et al.: Overcoming catastrophic forgetting in neural networks. Proc. Natl. Acad. Sci. **114**(13), 3521–3526 (2017)
10. Koutra, D., Tong, H., Lubensky, D.: Big-align: fast bipartite graph alignment. In: 2013 IEEE 13th International Conference on Data Mining. IEEE (2013)
11. Liang, S., Tang, S., Meng, Z., Zhang, Q.: Cross-temporal snapshot alignment for dynamic networks. IEEE Trans. Knowl. Data Eng. (2021)
12. Liu, L., Cheung, W.K., Li, X., Liao, L.: Aligning users across social networks using network embedding. In: IJCAI, pp. 1774–1780 (2016)
13. Man, T., Shen, H., Liu, S., Jin, X., Cheng, X.: Predict anchor links across social networks via an embedding approach. In: IJCAI, vol. 16, pp. 1823–1829 (2016)
14. Nguyen, T.T., Pham, M.T., Nguyen, T.T., Huynh, T.T., Nguyen, Q.V.H., Quan, T.T., et al.: Structural representation learning for network alignment with self-supervised anchor links. Expert Syst. Appl. **165**, 113857 (2021)
15. Ohsaka, N., Maehara, T., Kawarabayashi, K.i.: Efficient pagerank tracking in evolving networks. In: Proceedings of the 21th ACM SIGKDD International Conference on Knowledge Discovery and Data Mining, pp. 875–884 (2015)
16. Saxena, S., Chakraborty, R., Chandra, J.: HCNA: hyperbolic contrastive learning framework for self-supervised network alignment. Inf. Process. Manag. **59**(5), 103021 (2022)
17. Singh, R., Xu, J., Berger, B.: Global alignment of multiple protein interaction networks with application to functional orthology detection. Proc. Natl. Acad. Sci. **105**(35), 12763–12768 (2008)
18. Sun, L., Zhang, Z., Ji, P., Wen, J., Su, S., Philip, S.Y.: DNA: dynamic social network alignment. In: 2019 IEEE International Conference on Big Data (Big Data), pp. 1224–1231. IEEE (2019)
19. Sun, L., et al.: Aligning dynamic social networks: an optimization over dynamic graph autoencoder. IEEE Trans. Knowl. Data Eng. (2022)
20. Tang, J., Zhang, J., Yao, L., Li, J., Zhang, L., Su, Z.: ArnetMiner: extraction and mining of academic social networks. In: Proceedings of the 14th ACM SIGKDD International Conference on Knowledge Discovery and Data Mining, pp. 990–998 (2008)
21. Trung, H.T., Van Vinh, T., Tam, N.T., Yin, H., Weidlich, M., Hung, N.Q.V.: Adaptive network alignment with unsupervised and multi-order convolutional networks. In: 2020 ICDE, pp. 85–96. IEEE (2020)
22. Vijayan, V., Critchlow, D., Milenković, T.: Alignment of dynamic networks. Bioinformatics **33**(14), i180–i189 (2017)
23. Xiong, H., Yan, J., Pan, L.: Contrastive multi-view multiplex network embedding with applications to robust network alignment. In: Proceedings of the 27th ACM SIGKDD Conference on Knowledge Discovery & Data Mining (2021)
24. Yan, Y., Zhang, S., Tong, H.: BRIGHT: a bridging algorithm for network alignment. In: Proceedings of the Web Conference 2021, pp. 3907–3917 (2021)
25. Yang, L., et al.: HackGAN: harmonious cross-network mapping using CycleGAN with Wasserstein-Procrustes learning for unsupervised network alignment. IEEE Trans. Comput. Soc. Syst. (2022)
26. Zellinger, W., Grubinger, T., Lughofer, E., Natschläger, T., Saminger-Platz, S.: Central moment discrepancy (CMD) for domain-invariant representation learning. arXiv preprint arXiv:1702.08811 (2017)

27. Zhang, S., Tong, H.: Final: Fast attributed network alignment. In: Proceedings of the 22nd ACM SIGKDD International Conference on Knowledge Discovery and Data Mining, pp. 1345–1354 (2016)
28. Zhou, Y., et al.: Unsupervised adversarial network alignment with reinforcement learning. ACM TKDD **16**(3), 1–29 (2021)

CeDFormer: Community Enhanced Transformer for Dynamic Network Embedding

Jiaqi Guo[1], Tianpeng Li[2], Minglai Shao[2], Wenjun Wang[2,3]($\boxtimes$), Lin Pan[2], Xue Chen[2], and Yueheng Sun[2]

[1] School of Future Technology, Tianjin University, Tianjin, China
[2] College of Intelligence and Computing, Tianjin University, Tianjin, China
wjwang@tju.edu.cn
[3] Yazhou Bay Innovation Institute, Hainan Tropical Ocean University, Sanya Hainan 572022, China

Abstract. Dynamic Graph Representation Learning methods have achieved significant success. However, real-world data often do not strictly follow the smoothness assumption in time-dependent relationships between dynamic graph snapshots. Existing methods sometimes fail to capture these relationships effectively. To address this issue, we propose the Community Enhanced Transformer for Dynamic Network Embedding (CeDFormer), which leverages transformers to capture temporal structures in dynamic graphs. CeDFormer is designed to adapt to highly variable graph structures and complex temporal dependencies. To mitigate the high computational cost and limited scalability of transformers on large-scale graph data, we introduce an optimization strategy involving parameter sharing within stable communities from a global perspective. This strategy enhances training speed by $30\% \sim 35\%$ without compromising model performance. Extensive experiments on real-world datasets demonstrate that CeDFormer outperforms most other methods on the majority of datasets. Code is available at https://github.com/gjqwanttogjq/CeDFormer.

Keywords: Dynamic Network Embedding · Network Representation Learning · Dynamic Network Data

1 Introduction

In recent years, graph representation learning, also known as network embedding or graph embedding, has experienced rapid development. Its primary objective is to map nodes within a network into low-dimensional, real-valued, dense vectors while preserving network structure, node features, labels, and other auxiliary information [17,21,32]. Most existing graph representation learning methods rely on the assumption that the graph remains static. However, most real-world problems are better modeled as dynamic graphs, such as social networks in Facebook [30] and user-video interaction graphs in YouTube [5]. The structure and features of dynamic graphs change over time. While representation learning on static graphs has been studied extensively, dynamic graph representation learning is a relatively newer and more promising area of research. Introducing the temporal dimension adds complexity to the problem, making it more challenging, but also more applicable to real-world scenarios [31,33].

Existing methods have made several attempts to address the temporal dimension in dynamic graph analysis. For instance, approaches like VGRNN [13] have introduced recurrent neural networks (RNNs) to capture latent representations of nodes, while DGCN [10] employs LSTM to update GCN weight parameters, capturing global structural information across all snapshots in dynamic graphs. Others like DySAT [26], have introduced attention mechanisms to encode dynamic information that changes over time into node embeddings. DyFormer [8] builds upon attention mechanisms, incorporating transformers. Nonetheless, both of these methods still follow traditional approaches to handle the temporal dimension.

However, we have observed that real-world dynamic graph data do not always exhibit a strict chain-like dependency relationship,i.e., smoothness assumption, among snapshots over time. In other words, for a given snapshot, it may not necessarily have stronger dependencies on its adjacent snapshots. In the context of the real world, the evolution of each node does not necessarily occur within the time window between two consecutive snapshots [11,14]. If a set of nodes undergoes complex, non-smooth, non-uniform, and highly frequent changes over multiple snapshots, the application of recurrent neural networks (RNNs) would indeed pose challenges [16]. During the experimental phase, we have demonstrated that many datasets exhibit these characteristics. This observation aligns with reality; for instance, transportation networks evolve cyclically on a weekly basis [24], where the network on any given day is highly similar to the network from one week ago, rather than the network from the previous snapshot. Similarly, unexpected events in social networks can disrupt the assumption of network smoothness.

In response to the aforementioned challenges, we have developed a model based on the Transformer [28] architecture that is suitable for highly variable dynamic graph structures with intricate temporal dependencies between snapshots. To reduce overall complexity and training costs, we introduce an optimization strategy that involves parameter sharing within stable structural communities in dynamic graphs. Extensive experiments have confirmed that CeDFormer outperforms state-of-the-art methods on most real-world datasets, demonstrating the feasibility and effectiveness of our model and optimization strategy.

2 Related Work

2.1 Dynamic Graph Representation Learning

Previous methods for dynamic graph representation learning have typically extended static graph algorithms by introducing a temporal dimension to capture time-dependent relationships. Approaches for capturing temporal dependencies can be categorized into three main classes [8]:

Temporal Smoothness-Based Methods. These methods aim to ensure temporal smoothness in the embeddings of dynamic graph snapshots by incorporating temporal regularization. For example, DYNGEM [12] generates incremental embeddings of the current graph snapshot based on the embeddings of the previous snapshot. However, this approach heavily relies on the smoothness between snapshots and performs poorly when there are significant changes in node behavior between adjacent snapshots.

Recursive Methods. Recursive methods use recurrent neural networks (RNNs) to capture temporal dependencies. For instance, GCRN [27] combines convolutional neural networks (CNNs) and RNNs to simultaneously identify meaningful spatial structures and dynamic patterns in graph-structured data. Similarly, DGCN [10] uses Long Short-Term Memory (LSTM) [15] units to update GCN's weight parameters, capturing global structural information across all time steps. However, these methods have limitations when dealing with long-term dependencies and non-sequential dependencies, and they may lack scalability for large-scale dynamic graph data.

Attention-Based Methods. Attention mechanisms are used to aggregate spatial and temporal information. Traditional attention mechanisms, such as DySAT [26], aggregate information using self-attention mechanisms in both spatial and temporal dimensions. TGAT [31] encodes temporal information into node features and then applies attention mechanisms. However, traditional spatial attention mechanisms only consider attention calculations on existing edges, and temporal attention mechanisms focus solely on previous snapshots. In recent years, attention-based methods have seen significant development, especially with the introduction of transformers on graphs.

3 CeDFormer Model

3.1 Problem Definition

Suppose we have a dynamic graph $\mathcal{G} = \{G^{(1)}, G^{(2)}, ..., G^{(T)}\}$, where $G^{(t)} = (\mathcal{V}^{(t)}, \mathcal{E}^{(t)})$ represents a snapshot of the dynamic graph at time t. Each snapshot corresponds to a graph with a node set $\mathcal{V}^{(t)}$ and an edge set $\mathcal{E}^{(t)}$. There are no constraints between $G^{(t)}$ and $G^{(t+1)}$, which means that over time, between snapshots, we can expect changes in the number of nodes, the appearance of new edges, and the disappearance of existing edges [9,29]. CeDFormer takes variable-length sequences of adjacency matrices $\mathcal{A} = \{A^{(1)}, A^{(2)}, ..., A^{(T)}\}$ and node attribute matrices $\mathcal{X} = \{X^{(1)}, X^{(2)}, ..., X^{(T)}\}$ as input.

3.2 Method

In this section, we will introduce the architecture of the CeDFormer model (as illustrated in Fig. 1). The model consists of two main parts: the encoder and the decoder. Furthermore, we introduce a strategy for optimizing parameter count and training speed within the encoder module through community discovery methods.

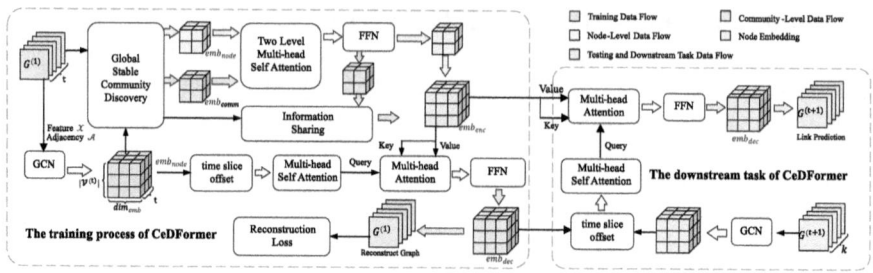

Fig. 1. The framework of the CeDFormer model, using link prediction as an example of downstream tasks, encompasses the learning process of the model from $G^{(1)}$ to $G^{(t)}$ and its application for link prediction from $G^{(t+1)}$ to $G^{(t+k)}$.

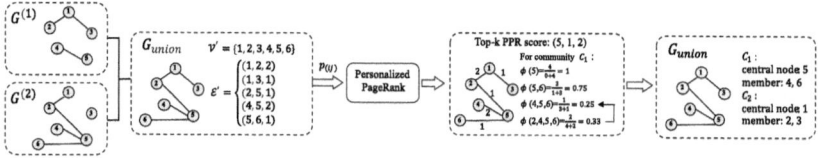

Fig. 2. An example of global stable community discovery, demonstrating how to perform local community discovery based on global influence $top-k$ nodes on $G^{(1)}$ and $G^{(2)}$.

Global Stable Community Discovery. Existing dynamic graph representation learning methods primarily focus on node-level embeddings, which pose scalability issues and computational inefficiencies, especially with large-scale graph data. To tackle this, we introduce a community perspective to detect stable and temporally invariant global community structures, illustrated in Fig. 2. These communities comprise nodes sharing similar structures and features, facilitating parameter sharing and enhancing model efficiency. Expanding on this, we present a strategy for local community discovery, centered on the top-k nodes with significant global influence.

Firstly, we aggregate the dynamic graph $\mathcal{G} = \{G^{(1)}, G^{(2)}, ..., G^{(T)}\}$ into a joint temporal graph $\mathcal{G}_{union} = (\mathcal{V}', \mathcal{E}')$. In this joint graph, $\mathcal{V}'$ contains all nodes that appear in any time snapshot, i.e., $\mathcal{V}' = \{i \mid \exists i \in \mathcal{V}^{(t)}, t \in T\}$. Similarly, $\mathcal{E}'$ consists of all edges present in any time snapshot, i.e., $\mathcal{E}' = \{(i,j) \mid \exists (i,j) \in \mathcal{E}^{(t)}, t \in T\}$. The weight of any edge (i,j) in the joint graph is defined as: $W'_{ij} = \sum_{t \in T} A^{(t)}_{ij}$. From a global perspective, our goal is to identify highly influential nodes to serve as central points for local community discovery. These nodes are likely to be the most influential within their respective communities. Global influence reflects a node's impact on maintaining the global structure, while intra-community influence reflects its ability to represent the entire community. To find the top-k nodes with the highest global influence, we perform random walks [23] on the joint graph $\mathcal{G}_{union}$. The probability $p(i,j)$ of transitioning from any node i to its neighboring node j is determined by the normalized weight of edge (i,j). This probability can be expressed as:

$$p(ij) = \frac{W'_{ij}}{\sum_{k \in \mathcal{N}(i)} W'_{ik}} \tag{1}$$

Here, $\mathcal{N}(i)$ represents the set of all neighboring nodes of node i.

Once we have calculated the transition probabilities between any two nodes in the joint graph, we compute the Personalized PageRank (PPR) [22] scores for all nodes on $\mathcal{G}_{union}$. The top-k nodes with the highest PPR scores are selected as central points for local community discovery [2]. These influential nodes serve as central points, forming the initial community structure. We then include all neighboring nodes within the community. Conductance [3] is used to measure the community structure, aiming to minimize inter-community edges while maximizing intra-community edges. For a community C, its conductance is defined as:

$$\phi(C) = \frac{\sum_{i \in C, k \in \overline{C}, (i,k) \in \mathcal{E}'} W'_{ik}}{\sum_{i,j \in C, (i,j) \in \mathcal{E}'} W'_{ij} + \sum_{i \in C, k \in \overline{C}, (i,k) \in \mathcal{E}'} W'_{ik}} \tag{2}$$

If attempting to pull in a certain node can reduce the conductance of a local community, it is considered to strengthen that local community, and we formally include the node in the local community. The community structures will be used in the encoder section of the model. In our design, we can gradually aggregate node information through multiple levels to achieve the transformation from the node level to the community level. The number of levels can be selected to strike a balance between performance and efficiency.

Embedding. In the encoder section, we start by aggregating the features of each snapshot of the dynamic graph using GCN (Graph Convolutional Network) [18]. For the snapshot $G^{(t)}$ at time t, we extract features, resulting in a feature matrix $\mathcal{F}_t = GCN(X^{(t)}, A^{(t)})$ of size $n \times dim_{emb}$, where n represents the number of nodes and dim_{emb} represents the node embedding dimensions. After aggregating features for all snapshots, we obtain a time-level node feature matrix of size $t \times n \times dim_{emb}$. This matrix is then used as input to the encoder layers.

To differentiate the temporal information between snapshots, we use position encoding in the transformer model to encode time information for each snapshot. The time dimension of the graph feature matrix enters the encoder layer and undergoes a dimension transformation, changing from a $t \times n \times dim_{emb}$ matrix to an $n \times t \times dim_{emb}$ matrix. This transformed matrix corresponds to the embeddings of all nodes at different time steps. Next, we optimize the model using the information from the global stable structural communities. For a community C, the central node i aggregates information from surrounding nodes using GCN. This process also aggregates the embeddings of other nodes within the community, propagating their information to the central node i. As a result, the embeddings of other nodes in community C can be replaced by the embedding of the central node $i : emb_{node}^i$ across all time steps. Based on this idea, we treat the embedding of the central node i as the embedding of community $C : emb_{comm}^C$, introducing a community-level encoder alongside the node-level encoder.

Encoder. Due to the fact that global community detection does not assign community membership to all nodes, there are still a few isolated nodes outside of all communities from a global perspective. These isolated nodes need to be embedded at the node level using an encoder-decoder structure. Both the embeddings of isolated nodes at each time step and community embeddings are used as inputs in the two hierarchical levels. Since the dimensions of embeddings are the same in both levels, the encoder structure does not need any additional modifications, meaning that the encoder structure is reusable between the two levels. In the encoder, we take the node embeddings emb_{node} and community embeddings emb_{comm}, both generated by the embedding module, as inputs for the node-level and community-level respectively. Due to the reusability of the encoder structure, we denote the inputs for both levels as emb_{enc}. emb_{enc} serves as the input for Key, Query, and Value in the multi-head attention layer, allowing multiple heads to create distinct subspaces for the model to focus on different aspects of information.

In the encoder structure, introducing a community-level perspective can indeed lead to significant savings in training time. Assuming that during the phase of discovering stable structural communities at a global level, you find that there are c communities, each with m members, and the number of isolated nodes is $|\mathcal{V}'| - c * m$. Then, compared to not using the optimization strategy, the overall optimization rate for the entire encoder part can be calculated as $\frac{c*(m-1)}{|\mathcal{V}'|}$.

Decoder. In the decoder part, we start again from the node perspective to compute embeddings for each node. Additionally, the process in the decoder layer differs between the training phase and the downstream task phase.

In the decoder part, we approach the problem from a node-centric perspective, calculating embeddings for each node. Furthermore, the decoder layer has two multi-head attention layers.

The first one is similar to the multi-head attention layer in the encoder, but we have made two improvements. Firstly, we introduce a concept similar to the "Beginning of Sentence" (BOS) token used in natural language processing (NLP). In CeFormer, the concept of the BOS token is specifically manifested in the initial graph embedding. During the training process, the initial graph embedding is placed before the first time snapshot. It is obtained through random initialization during the training phase and, in the downstream task phase, it is derived from the encoder's output in the training phase. Additionally, we remove the last time snapshot to maintain a consistent input size. This approach is designed to ensure that the training data at the current moment does not affect the calculation of node embeddings at the current time snapshot. Secondly, we have also introduced an additional sequence mask mechanism in the decoder, similar to the sequence mask used in transformers. The mask mechanism ensures that the model relies only on information from previous snapshots during node embedding calculations. Through these two methods, when reconstructing any snapshot at time t, we rely only on information from snapshots $0, ..., t-1$ (for t

> 1). The graph at time 0 is reconstructed from learnable, randomly initialized initial graph embeddings.

The second multi-head attention layer takes the output of the first one as a query and the output of the encoder to recalculate the node embedding matrix emb'_{enc}, with dimensions of $n \times t \times dim_{emb}$, which is used as key and value. The embedding of any node i within emb'_{enc}, denoted as $emb'_{enc}i$, can be represented as follows:

$$emb'_{enc}i = \begin{cases} emb'_{node}i & \nexists m, \text{ s.t. } i \in m \\ emb'_{comm}m & \exists m, \text{ s.t. } i \in m \end{cases} \quad (3)$$

Similar to the encoder, the output of the second multi-head attention layer in the decoder undergoes normalization and a feedforward layer to obtain embeddings for each node across all time steps. These embeddings are then combined, transformed to the matrix emb_{dec}. After passing through a linear layer, emb_{dec} is further transformed into reconstruction snapshots of dynamic graphs with dimensions $n \times n$.

Training. The model's objective function is derived from the reconstruction error for each snapshot. The reconstruction error $Loss_t$ is computed using the following formula:

$$Loss_t = -\frac{1}{|\mathcal{V}^{(t)}|^2} \sum_{i=1}^{|\mathcal{V}^{(t)}|} \sum_{j=1}^{|\mathcal{V}^{(t)}|} adj(t)_{ij} \cdot \log(adj_{rec}(t)_{ij}) \\ + (1 - adj(t)_{ij}) \cdot \log(1 - adj_{rec}(t)_{ij}) \quad (4)$$

Here, $adj(t)_{i,j}$ represents the value of the edge i,j in the original adjacency matrix at time t, and $adj_{rec}(t)_{i,j}$ represents the corresponding value of the edge i,j in the reconstructed snapshot at time t. The $adj_{rec}(t)$ values are obtained by passing the embeddings at time t, $emb_{dec}(t)$, through a linear layer.

The overall loss for the entire model is obtained by summing up the losses for each time step, represented as:

$$Loss = \sum_{t=1}^{T} Loss_t \quad (5)$$

During the model training phase, the encoder learns to obtain better node representations, while the decoder learns how to effectively utilize node representations and historical information to reconstruct the dynamic graph.

3.3 Model Complexity Analysis

For a dynamic graph with a time slice length of t, maximum nodes v, maximum edges e, embedding dimension d, transformer layers n, and attention heads h, neglecting global community optimization, CeDFormer's time complexity is $O(etd + nvt^2d)$, and space complexity is $O(hnvd^2t^2)$. CeDFormer's design involves applying transformers at each node, thus scaling with increasing nodes without causing an unacceptable decline in model efficiency.

4 Experiments

4.1 Experimental Setup

Dataset and Baseline Models. To demonstrate CeDFormer's performance, we conducted an in-depth study of standard datasets from related literature. We selected five widely-used datasets: enron10 [6], DBLP [13], highSchool [10], Facebook [25], Email [10] and bitcoinOTC [10]. These datasets were used for tasks such as link prediction, clustering, and time efficiency comparisons. We also analyzed the temporal dependencies in these datasets. Our baseline models included: VGRNN [13], DySAT [26], DyFormer [8], DGCN [10], HGWaveNet [4]. We followed the settings described in the original papers of these baseline models. The specific number of nodes, edges, the number of snapshots, and the number of test snapshots for each dataset are listed in Table 1.

Table 1. Dataset description

Dataset	Nodes	Edges	Snapshots	Test Snapshots
Enron	184	4784	11	3
DBLP	315	5104	10	3
highSchool	327	26870	9	3
Facebook	663	23394	9	3
Email	2029	28090	29	3
bitcoinOTC	5053	11651	12	3

Evaluation Metrics. In the time snapshot dependency analysis section, for a dynamic graph $\mathcal{G} = \{G^{(1)}, G^{(2)}, ..., G^{(T)}\}$, where $t > 1$, we calculate the cosine similarity between snapshot $G^{(t)}$ and all the previous time snapshots $G^{(1)}, G^{(2)}, ...G^{(t-1)}$. This calculation is used to measure the dependency relationship between any two time snapshots. The calculation of cosine similarity involves reducing the dimensionality of the two snapshots and then computing the cosine similarity on two one-dimensional vectors.

In the context of dynamic graphs with partially observed snapshots $\mathcal{G} = \{G^{(1)}, G^{(2)}, ..., G^{(T)}\}$, link prediction tasks can be described as follows:

1) Link Prediction: This involves predicting the edges in $G^{(t+1)}$, which means forecasting which edges will exist in the next snapshot.
2) New Link Prediction: This task specifically focuses on predicting edges in $G^{(t+1)}$ that are not present in $G^{(t)}$.

Hyper-parameters. The link prediction experiments follow the same settings as in VGRNN. We randomly select 5% and 10% of the edges from each snapshot as the validation and test sets for the link prediction task. The embedding dimension for all tasks was set to 32. In the experiments, CeDFormer defaults to using all nodes for the global stable community optimization strategy. All experiments were conducted using a single RTX3090 GPU with 24GB of VRAM.

4.2 Experiment Results

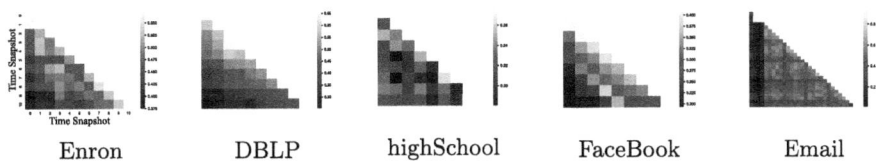

Fig. 3. Time Snapshot Dependency Analysis

Time Snapshot Dependency Analysis. Based on Fig. 3, we conducted dependency analysis on the mentioned datasets. It can be observed that the diagonal blocks in the dependency matrices for the Enron and DBLP data have lighter colors, corresponding to cosine similarities exceeding 0.5. This indicates stronger dependencies between each snapshot and the previous one, aligning these datasets more closely with traditional time series data and adhering more to the Markov assumption: that the conditional probability distribution of future states depends only on the current state. In particular, DBLP data highlights the transitivity of time-dependent relationships, meaning that the similarity between any two snapshots is influenced by the time gap between them.

Similarly, the Facebook data also exhibits the transitivity of dependencies, but the snapshot dependencies within the Facebook dataset itself are not very pronounced, with diagonal elements in the matrix representing cosine similarities not exceeding 0.4.

For the highSchool data, it does not exhibit clear dependencies, especially in snapshots 4 and 6, where it shows highly inconsistent characteristics compared to the previous snapshot. Such data is more variable, displays weaker time dependencies, and tends to exhibit dependencies across snapshots.

In the case of the Email dataset, we observed a high degree of change and periodic variations. This dataset undergoes significant changes from snapshot 3 to 4, resulting in minimal dependencies between later snapshots and snapshots 0 to 3.

As designed, CeDFormer performs better on data resembling highSchool data, leading to more significant improvements compared to baseline methods.

Link Prediction. The experimental results for link prediction on various datasets are presented in Table 2. From the table, we can observe that CeDFormer performs comparably to HGWaveNet on the Enron dataset and is approximately 1.3% behind HGWaveNet on the DBLP dataset. However, on the highSchool, bitcoinOTC, Facebook, and Email datasets, CeDFormer outperforms the latest baselines, indicating that CeDFormer performs better on highly variable and periodic data, as supported by the analysis of time snapshot dependencies.

Table 2. AUC and AP scores of link prediction. The best are bolded and the second best are underlined.

Metrics	Model	Enron	DBLP	highSchool	FaceBook	Email	bitcoinOTC
AUC	VGRNN	93.42±0.70	85.80±0.78	89.66±0.32	<u>89.79±0.34</u>	91.92±1.18	81.34±3.25
	DySAT	88.81±1.10	86.74±1.51	91.51±0.48	88.88±0.89	90.42±0.91	83.06±2.71
	Dyformer	90.35±0.45	77.74±0.63	85.01±0.29	83.74±0.76	91.08±0.44	83.62±1.01
	DGCN	85.27±0.85	72.03±0.54	66.18±0.53	68.65±0.29	<u>95.12±0.30</u>	84.35±4.88
	HGWaveNet	<u>94.45±0.26</u>	**88.95±0.47**	<u>91.64±0.22</u>	86.98±0.66	92.49±0.36	80.06±1.27
	CeDformer	**94.67±0.51**	<u>87.64±0.92</u>	**94.33±0.80**	**90.05±0.50**	**95.83±1.01**	**89.23±1.97**
AP	VGRNN	**94.50±0.66**	88.64±0.58	88.71±0.65	<u>89.14±0.41</u>	93.34±0.70	<u>88.62±2.37</u>
	DySAT	87.30±1.54	89.64±1.00	89.17±1.12	88.61±0.96	89.19±1.17	81.18±2.59
	Dyformer	90.78±0.44	78.94±0.51	84.25±0.34	81.16±0.72	92.39±0.49	84.54±1.74
	DGCN	84.51±0.79	73.14±0.69	66.09±0.56	68.78±0.34	<u>95.32±0.31</u>	81.42±3.74
	HGWaveNet	<u>94.37±0.32</u>	**91.72±0.38**	<u>90.86±0.25</u>	85.96±0.71	94.02±0.32	81.15±1.24
	CeDformer	93.87±0.39	<u>90.52±0.48</u>	**94.10±0.83**	**89.84±0.56**	**96.65±0.42**	**88.63±2.70**

Table 3. The average training time for one epoch on different datasets, in seconds/epoch.

Model	Enron	DBLP	highSchool	FaceBook	Email
DySAT	2.240	2.268	2.740	4.622	8.620
Dyformer	1.123	2.474	4.728	6.822	17.774
HGWaveNet	67.740	95.316	127.335	262.109	683.212
CeDFormer	2.267	3.319	2.945	5.972	25.719

The Table 3 illustrates the average runtime per epoch on the dataset for each method (measured in seconds) for link prediction. We compare the efficiency of CeDFormer with other deep baseline methods. Consistent with our hypothesis,

Table 4. AUC and AP scores of new link prediction. The best are bolded and the second best are underlined.

Metrics	Model	Enron	DBLP	highSchool	FaceBook	Email	bitcoinOTC
AUC	VGRNN	87.41±0.96	75.87±1.65	88.09±0.28	<u>86.76±0.54</u>	90.37±1.28	80.22±2.78
	DySAT	82.58±3.00	76.57±2.30	<u>90.42±0.78</u>	85.97±1.21	81.71±3.97	83.09±2.61
	Dyformer	87.35±0.52	74.76±0.66	84.23±0.33	81.58±0.69	90.62±0.68	81.31±1.69
	DGCN	81.87±0.70	62.48±1.54	62.08±0.28	64.98±0.43	<u>93.21±0.40</u>	83.50±1.83
	HGWaveNet	<u>89.38±0.36</u>	**83.73±0.55**	89.63±0.24	83.97±0.61	92.05±0.38	79.97±0.99
	CeDformer	**90.46±0.59**	<u>80.41±0.88</u>	**91.35±0.78**	**88.37±0.46**	**94.36±0.49**	**88.72±1.55**
AP	VGRNN	88.76±0.70	78.38±1.21	86.81±0.54	85.30±0.66	92.22±0.77	<u>84.85±3.45</u>
	DySAT	83.07±2.89	78.81±2.13	<u>88.76±1.56</u>	84.78±1.79	74.96±3.13	82.24±2.94
	Dyformer	88.00±0.46	73.96±0.73	84.25±0.31	80.66±0.55	91.24±0.76	80.02±2.06
	DGCN	82.03±0.57	64.72±1.52	62.15±0.37	65.65±0.53	<u>93.36±0.43</u>	81.38±1.15
	HGWaveNet	87.71±0.33	**86.89±0.47**	88.48±0.27	82.30±0.67	93.29±0.38	79.86±1.25
	CeDformer	**89.74±0.71**	<u>83.39±0.64</u>	**90.09±0.72**	**87.91±0.45**	**95.75±0.53**	**87.11±2.01**

despite utilizing a transformer-based architecture, through optimization techniques, the efficiency of our model is comparable to existing methods and significantly outperforms the HGWaveNet model on hyperbolic space.

New Link Prediction. New link prediction is used to predict new edges that will appear in the next snapshot, evaluating the model's inductive ability. This task is more challenging, and both link prediction and new link prediction tasks are performed in the same model training process. Compared to link prediction tasks, the performance of new link prediction tasks may experience a certain degree of decline. However, as shown in Table 4, CeDFormer demonstrates better generalization on the Enron dataset, surpassing HGWaveNet and achieving better results. Furthermore, it maintains a leading position on the highSchool, bitcoinOTC, Facebook, and Email datasets. Compared to other self-attention-based methods or transformer-based methods, we harness the attention mechanism's performance more effectively on dynamic graph data.

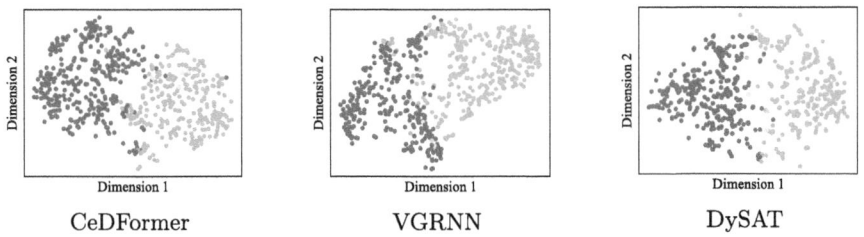

Fig. 4. The results of node representation analysis, t-SNE Visualization of Graph Embeddings (Color figure online)

4.3 Node Representation Analysis

In this experiment, we used t-SNE (t-Distributed Stochastic Neighbor Embedding) [20] to reduce the node embeddings of the Facebook dataset snapshots to two dimensions for visualization. We employed the same t-SNE parameters for each model, setting the cluster density ($perplexity$) to 10 and the maximum number of iterations(n_iter) to 300.

From Fig. 4, it can be observed that although CeDFormer requires more complex methods like calculating attention and using linear layers in the decoder module to reconstruct snapshots for node embeddings, the quality of node embeddings doesn't deteriorate. Compared to VGRNN, CeDFormer's t-SNE visualization shows smaller areas of intersection between two classes of labels (red and blue), and it is on par with DySAT based on self-attention mechanisms.

4.4 Ablation Study

To further validate the reliability of our model's global community optimization strategy, we conducted ablation experiments on the relevant datasets. We also

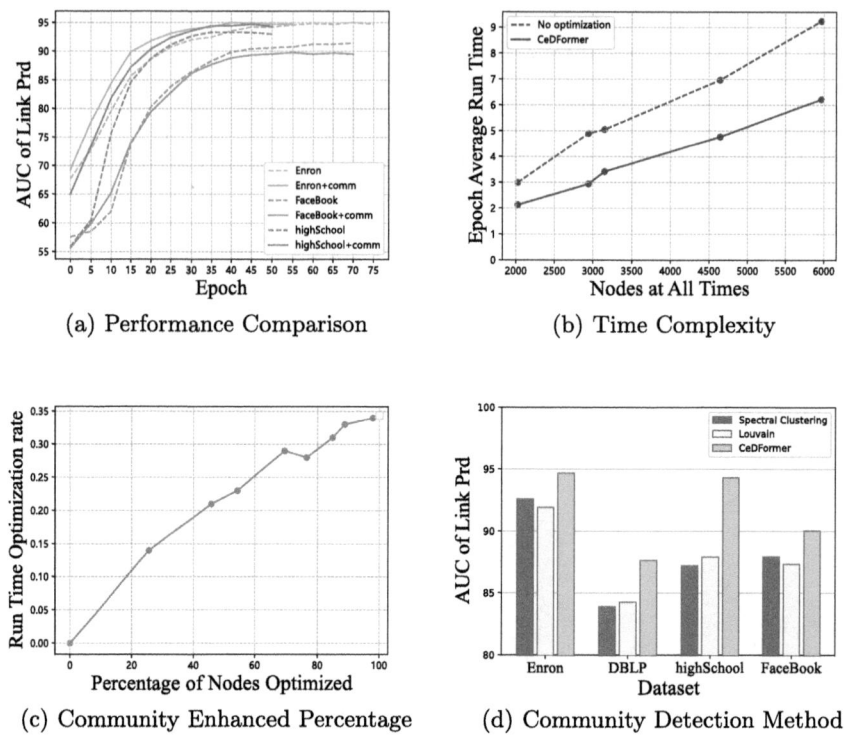

Fig. 5. The results of ablation experiment (Color figure online)

explored the impact of the proportion of nodes in community levels and different community discovery methods on optimization performance and efficiency.

Performance Comparison. As shown in Fig. 5(a), we conducted experiments on the Enron, Facebook, and highSchool datasets. Different colored lines represent different datasets. The solid lines represent CeDFormer, while the dashed lines represent the model without community-level optimization. There is no significant difference in performance before and after removing the optimization strategy. CeDFormer converges faster on the Enron and highSchool datasets, with around a 1.8% performance improvement on the highSchool dataset. However, after removing the global stable community structure optimization strategy, CeDFormer's performance decreases by around 1.6% on the Facebook dataset.

Time Efficiency Analysis. As shown in Fig. 5(b), we compared the running time of the model per epoch before and after removing the optimization strategy at different node counts (where the total node count is obtained by adding the number of nodes from all snapshots). It can be observed that as the node count increases, using the optimization strategy still reduces training time by

30% ~ 35% . The optimization effect is more significant on datasets with a more pronounced community structure.

Analysis of the Proportion of Community-Level Nodes. The global stable structure optimization strategy allows us to choose the proportion of nodes participating in the community level, relative to all nodes. As shown in Fig. 5(c), when this proportion is set to 0, the CeDFormer reverts to having only the node-level. The proportion of community-level nodes starts from 25.6%, given that the initial selection of nodes generates larger local communities. It can be observed that the optimization rate does not increase linearly with this proportion. This is because the process of selecting central nodes is sorted by influence, and nodes selected later have less influence, resulting in smaller local community structures and less effective optimization through parameter sharing.

Community Detection Methods Analysis. As shown in Fig. 5(d), we conducted a performance analysis by replacing our global influence-based local community detection method on four datasets. It can be observed that the Louvain algorithm [7] and spectral clustering methods [1,19] did not perform well in the context of joint graph community detection for global stability. In particular, the Louvain algorithm had excessive time overhead on large-scale graph data and is not a viable choice. This experiment further validates the effectiveness of our optimization strategy.

Table 5. AUC and AP scores of link prediction on the BitcoinOTC Dataset.

Model	bitcoinOTC		bitcoinOTC-4k		bitcoinOTC-3k		bitcoinOTC-2k		bitcoinOTC-1k	
	AUC	AP	AUC	AP	AUC	AP	AUC	AP	AUC	AP
VGRNN	81.34	88.62	82.02	89.31	81.19	89.05	76.95	85.99	63.07	71.40
DySAT	83.06	81.18	86.32	85.20	84.19	81.29	87.37	85.90	68.54	74.31
CeDFormer	89.23	88.63	93.48	94.05	93.22	95.03	88.42	85.21	73.01	75.20

4.5 Model Robustness Experiment

Node Quantity Robustness Experiment. We partitioned the bitcoinOTC data into subsets containing 1000, 2000, 3000, and 4000 nodes. We conducted experiments to assess CeDFormer's robustness with increasing node counts across these datasets, using VGRNN and DySAT as baselines. The specific experimental results are presented in Table 5. Observing the results, CeDFormer outperformed both VGRNN and DySAT across all versions of the bitcoinOTC dataset. It is worth noting that CeDFormer demonstrates excellent performance on datasets with 2k and 3k nodes, indicating that the model can effectively reconstruct missing information when some information is absent. However, on the 1k-node dataset, due to its sparse nature and significant information loss, the model's performance may decline, yet it still outperforms VGRNN and DySAT.

As illustrated in Fig. 6, We plotted the average training time per epoch related to the increase in the node count on the bitcoinOTC dataset. It is evident that the overall runtime of CeDFormer is almost linearly correlated with the number of nodes, consistent with the results of our previous time complexity analysis.

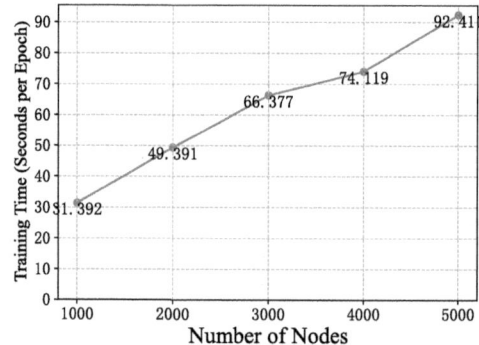

Fig. 6. The training time per epoch (seconds/epoch) as a function of the number of nodes

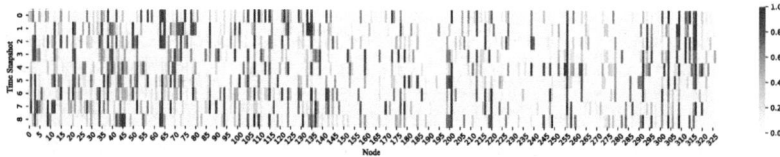

Fig. 7. Analysis of influential nodes in the highSchool Dataset

Robustness Experiment of Influential Nodes. We replicated the method used in CeDFormer to select influential nodes. This process was executed in each snapshot, extracting the top 100 influential nodes and comparing their distributions across all snapshots. The visualization of node influence for each snapshot is depicted in Fig. 7, where the heatmap rows represent individual snapshots, columns represent nodes, and color intensity indicates the node's influence within a snapshot. Darker colors denote higher influence. The analysis revealed that the distribution of the top 100 influential nodes across each snapshot in the highSchool dataset appeared disordered. There was significant variation in influential nodes across snapshots.

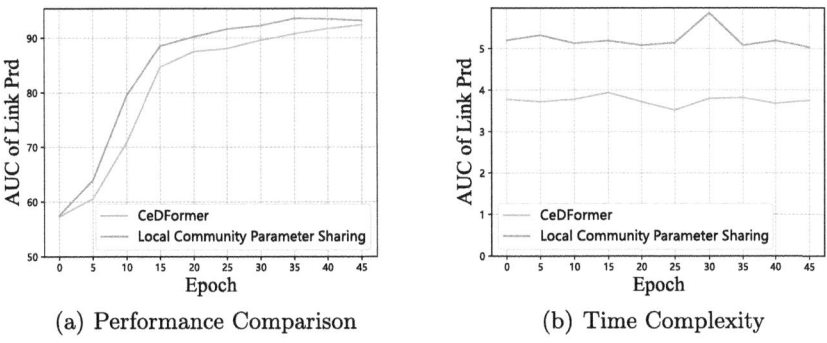

Fig. 8. The results of ablation experiment

Then we replaced CeDFormer's global community optimization strategy with a strategy involving parameter sharing within communities in each snapshot. This approach strictly identified each influential node, even in cases of variability. It's important to note that this strategy did not identify stable global communities, eliminating the efficiency optimization introduced by CeDFormer's community hierarchy, and only reduced the parameter count due to parameter sharing. Figure 8 illustrates the effectiveness and efficiency of this approach. Even in datasets with highly variable influential nodes, CeDFormer performs slightly less than the local parameter sharing strategy for influential nodes in time slices, but it achieves optimization in efficiency, remaining consistent with the original design motivation of the model.

5 Conclusion

In this paper, we introduced CeDFormer, a model designed to adapt better to highly variable graph structures and complex dependencies between snapshots. We proposed an optimization strategy that involves sharing parameters within stable communities from a global perspective on the graph structure. Extensive experiments demonstrate the effectiveness and versatility of our proposed framework compared to state-of-the-art methods. In future experiments, we will progressively aggregate node information through a bottom-up, multi-level community detection approach, aiming to transform node-level information into community-level representation. We will also explore methods and metrics to select the optimal number of community levels to balance performance and efficiency.

Acknowledge. This work was supported in part by Key R&D Project of Hainan (Grant No.ZDYF 2024SHFZ051), National Key R&D Plan of China (Grant No.2022YFC2602305) and ENNGroup under Grant 2023GKF-1220.

References

1. Amini, A.A., Chen, A., Bickel, P.J., Levina, E.: Pseudo-likelihood methods for community detection in large sparse networks (2013)
2. Andersen, R., Chung, F.: Detecting sharp drops in pagerank and a simplified local partitioning algorithm. In: International Conference on Theory and Applications of Models of Computation, pp. 1–12. Springer (2007)
3. Andersen, R., Chung, F., Lang, K.: Local graph partitioning using pagerank vectors. In: 2006 47th Annual IEEE Symposium on Foundations of Computer Science (FOCS'06), pp. 475–486. IEEE (2006)
4. Bai, Q., Nie, C., Zhang, H., Zhao, D., Yuan, X.: HGWaveNet: a hyperbolic graph neural network for temporal link prediction. In: Proceedings of the ACM Web Conference 2023, pp. 523–532 (2023)
5. Benevenuto, F., Duarte, F., Rodrigues, T., Almeida, V.A., Almeida, J.M., Ross, K.W.: Understanding video interactions in YouTube. In: Proceedings of the 16th ACM International Conference on Multimedia, pp. 761–764 (2008)
6. Benson, A.R., Abebe, R., Schaub, M.T., Jadbabaie, A., Kleinberg, J.: Simplicial closure and higher-order link prediction. Proc. Natl. Acad. Sci. **115**(48), E11221–E11230 (2018)
7. Blondel, V.D., Guillaume, J.L., Lambiotte, R., Lefebvre, E.: Fast unfolding of communities in large networks. J. Stat. Mech. Theory Exp. **2008**(10), P10008 (2008)
8. Cong, W., et al.: DyFormer: a scalable dynamic graph transformer with provable benefits on generalization ability. In: Proceedings of the 2023 SIAM International Conference on Data Mining (SDM), pp. 442–450. SIAM (2023)
9. Gao, C., Fan, Y., Jiang, S., Deng, Y., Liu, J., Li, X.: Dynamic robustness analysis of a two-layer rail transit network model. IEEE Trans. Intell. Transp. Syst. **23**(7), 6509–6524 (2021)
10. Gao, C., Zhu, J., Zhang, F., Wang, Z., Li, X.: A novel representation learning for dynamic graphs based on graph convolutional networks. IEEE Trans. Cybern. (2022)
11. Goyal, P., Chhetri, S.R., Canedo, A.: dyngraph2vec: capturing network dynamics using dynamic graph representation learning. Knowl.-Based Syst. **187**, 104816 (2020)
12. Goyal, P., Kamra, N., He, X., Liu, Y.: DynGEM: deep embedding method for dynamic graphs. arXiv preprint arXiv:1805.11273 (2018)
13. Hajiramezanali, E., Hasanzadeh, A., Narayanan, K., Duffield, N., Zhou, M., Qian, X.: Variational graph recurrent neural networks. Adv. Neural Inf. Process. Syst. **32** (2019)
14. Harary, F., Gupta, G.: Dynamic graph models. Math. Comput. Model. **25**(7), 79–87 (1997)
15. Hochreiter, S., Schmidhuber, J.: Long short-term memory. Neural Comput. **9**(8), 1735–1780 (1997)
16. Karita, S., et al.: A comparative study on transformer vs RNN in speech applications. In: 2019 IEEE Automatic Speech Recognition and Understanding Workshop (ASRU), pp. 449–456. IEEE (2019)
17. Kazemi, S.M., et al.: Representation learning for dynamic graphs: a survey. J. Mach. Learn. Res. **21**(1), 2648–2720 (2020)
18. Kipf, T.N., Welling, M.: Semi-supervised classification with graph convolutional networks. arXiv preprint arXiv:1609.02907 (2016)

19. de Lange, S.C., de Reus, M.A., van den Heuvel, M.P.: The Laplacian spectrum of neural networks. Front. Comput. Neurosci. **7**, 189 (2014)
20. Linderman, G.C., Rachh, M., Hoskins, J.G., Steinerberger, S., Kluger, Y.: Efficient algorithms for t-distributed stochastic neighborhood embedding. arXiv preprint arXiv:1712.09005 (2017)
21. Liu, X., Tang, J.: Network representation learning: a macro and micro view. AI Open **2**, 43–64 (2021)
22. Page, L.: The pagerank citation ranking: bringing order to the web, Technical report, Stanford Digital Library Technologies Project, 1998 (1998)
23. Perozzi, B., Al-Rfou, R., Skiena, S.: DeepWalk: online learning of social representations. In: Proceedings of the 20th ACM SIGKDD International Conference on Knowledge Discovery and Data Mining, pp. 701–710 (2014)
24. Rahimi, A., Cohn, T., Baldwin, T.: Semi-supervised user geolocation via graph convolutional networks. arXiv preprint arXiv:1804.08049 (2018)
25. Rossi, R., Ahmed, N.: The network data repository with interactive graph analytics and visualization. In: Proceedings of the AAAI Conference on Artificial Intelligence, vol. 29 (2015)
26. Sankar, A., Wu, Y., Gou, L., Zhang, W., Yang, H.: DySAT: deep neural representation learning on dynamic graphs via self-attention networks. In: Proceedings of the 13th International Conference on Web Search and Data Mining, pp. 519–527 (2020)
27. Seo, Y., Defferrard, M., Vandergheynst, P., Bresson, X.: Structured sequence modeling with graph convolutional recurrent networks. In: Cheng, L., Leung, A.C.S., Ozawa, S. (eds.) ICONIP 2018. LNCS, vol. 11301, pp. 362–373. Springer, Cham (2018). https://doi.org/10.1007/978-3-030-04167-0_33
28. Vaswani, A., et al.: Attention is all you need. Adv. Neural Inf. Process. Syst. **30** (2017)
29. Wang, Z., Wang, C., Gao, C., Li, X., Li, X.: An evolutionary autoencoder for dynamic community detection. Sci. Chin. Inf. Sci. **63**(11), 1–16 (2020). https://doi.org/10.1007/s11432-020-2827-9
30. Weaver, J., Tarjan, P.: Facebook linked data via the graph API. Seman. Web **4**(3), 245–250 (2013)
31. Xu, D., Ruan, C., Korpeoglu, E., Kumar, S., Achan, K.: Inductive representation learning on temporal graphs. arXiv preprint arXiv:2002.07962 (2020)
32. Zhu, Y., et al.: A survey on graph structure learning: progress and opportunities. arXiv preprint arXiv:2103.03036 (2021)
33. Zhu, Y., Lyu, F., Hu, C., Chen, X., Liu, X.: Encoder-decoder architecture for supervised dynamic graph learning: a survey. arXiv preprint arXiv:2203.10480 (2022)

Challenges and Solutions in Drift Detection and Anomaly Explanation

On the Impact of Industrial Delays when Mitigating Distribution Drifts: An Empirical Study on Real-World Financial Systems

Thibault Simonetto[1]([✉]), Maxime Cordy[1], Salah Ghamizi[2],
Yves Le Traon[1], Clément Lefebvre[3], Andrey Boystov[3], and Anne Goujon[3]

[1] University of Luxembourg, Luxembourg, Luxembourg
{thibault.simonetto,maxime.cordy,yves.letraon}@uni.lu
[2] Luxembourg Institute of Science and Technology, Luxembourg, Luxembourg
salah.ghamizi@list.lu
[3] BGL BNP Paribas, Paris, France
{andrey.boytsov,anne.goujon}@bgl.lu

Abstract. An increasing number of financial software system relies on Machine learning models to support human decision-makers. Although these models have shown satisfactory performance to support human decision-makers in classifying financial transactions, the maintenance of such ML systems remains a challenge. After deployment in production, the performance of the models tends to degrade over time due to concept drift. Methods have been proposed to detect concept drift and retrain new models upon detection to mitigate the drop in performance. However, little is known about the effectiveness of such methods in an industrial context. In particular, their evaluation fails to consider the delay between the detection of the drift and the deployment of a new model. This delay is inherent to the strict quality assurance and manual validation processes that financial (and other critical) institutions impose on their software systems. To circumvent this limitation, we formalize the problem of retraining ML models against distribution drift in the presence of delay and propose a novel protocol to evaluate drift detectors. We report on an empirical study conducted on the transaction system of our industrial partner, BGL BNP Paribas, and two publicly available datasets: Lending Club Loan Data and Electricity. We release our tool and benchmark on GitHub. [1] We demonstrate for the first time how ignoring the delays in the evaluation of the drift detectors overestimates their ability to mitigate performance drift, up to 39.86% for our industrial application.[1] Code available at https://github.com/serval-uni-lu/drift-robustness.

Keywords: ML · distribution-drift · real-world system · AI in finance

1 Introduction

Machine Learning (ML) is used in industry to leverage an increasingly large amount of data and reduce operational costs, develop new disruptive products, and deliver personalized services to their customers.

Our industrial partner is an important actor in the financial sector, and has many operational benefits from automated services built using ML, including scalability at low costs and the ability to process large amounts of data efficiently.

However, the wide dissemination of ML technologies within industrial software systems is hindered by their high maintenance costs [1]. Our partner has observed that the effectiveness (e.g. prediction accuracy) of their ML systems declines over time due to changes in data distribution.

The usual solution to mitigate the effect of drifts on ML systems is to retrain the ML model periodically (periodic retraining) or continuously (online learning). In our partner's case, online learning is prohibited by stringent security policies, which prevent feeding models with live data without manual checks. Thus, our partner relies on periodic retraining. The critical questions they face are *how often* and *how* they should retrain their model.

Alternatively, research has developed *drift detectors* as a better means to decide when to trigger model retraining [14]. A drift detector is a statistics-based method that takes as input a stream of samples. After each sample, it returns whether the observed distribution has shifted or not compared to previous samples. Using drift detectors to trigger retraining at the most appropriate times can reduce the cost of periodic retraining and increase its effectiveness.

Although the use of periodic retraining and drift detectors has been intensively investigated in the literature [3,14,18], previous studies do not consider the industrial constraints facing ML systems in production. In particular, two types of delay inhibit the retraining process. First, *labeling delay* implies a temporal distance between the time at which a new data reaches the system and the time at which its ground truth label is retrieved. Second, *deployment delay* is inherent to the strict quality assurance and manual validation processes that financial institutions (and other critical institutions) impose on their software systems. Hence, there is a significant time gap between when a system is fully engineered (or updated) and when it is running in production. In the case of BGL BNP Paribas, this delay is typically 10 days for labeling, and 28 days for deployment.

In this paper, our objective is to uncover the capabilities of retraining strategies to mitigate the effect of drifts *in presence* of labeling and deployment delays. Therefore, we conduct an empirical study involving one real-world financial system of our partner, and one publicly available dataset. We cover 16 retraining scheduling methods (based on periodic retraining or drift detectors). We measure the capability of these strategies to retrain models efficiently (minimizing the number of retraining) and effectively (maximizing performance over time). We specifically investigate the impact that the aforementioned delay has on existing retraining practices. To summarize, our study brings three novel contributions:

- We formulate the problem of retraining against distribution drifts in presence of *labeling and deployment delays*.
- We propose a novel step-by-step protocol for ML practitioners to diagnose concept drifts with deployment delays and identify the best retraining strategies that fit their case.
- We report on an empirical study of the effectiveness and efficiency of retraining strategies. We notably shed light on the impact of window size of retraining, the importance of drift detectors tuning, and how the delay affects the Pareto-optimal retraining strategies.

Through our study, we highlight the importance that labeling and deployment delays have on the predictive maintenance of machine learning-based systems, and the scale at which these delays impact the solutions to combat distribution drifts in the real world. By providing a proper definition and evaluation protocol of this industry-relevant problem overlooked by the literature, we hope to inspire future research on designing effective and efficient solutions.

2 Background

2.1 Source of Performance Drift

In their survey [14], Lu et al. define concept drift (also defined as distribution shift [22]) as follows:

Given a time period $[0, t]$, the set of samples, denoted $S_{0,t} = \{d_0, \ldots, d_t\}$, where $d_i = (x_i, y_i)$ is one observation (or a data instance), x_i is the feature vector, y_i is the label and $S_{0,t}$ follows a certain distribution $F_{0,t}(X, y)$. Concept drift occurs at timestamp $t + 1$, if $F_{0,t}(X, y) \neq F_{t+1,\infty}(X, y)$, denoted $\exists t : P_t(X, y) \neq P_{t+1}(X, y)$.

According to this definition, concept drift can be defined as the change in the joint probability of X and y at time t. The joint probability $P_t(X, y)$ can be decomposed as $P_t(X, y) = P_t(X) \times P_t(y|X)$, therefore, the concept drift can have three sources. We describe the three sources of concept drift and explain how each source can influence the performance of the model.

- $P_t(X) \neq P_{t+1}(X)$ while $P_t(y|X) = P_{t+1}(y|X)$, that is, only the input space distribution $P_t(X)$ changes and the relation $P_t(y|X)$ remains unchanged. This kind of drift does not affect the true decision boundary. Therefore, it may or may not affect the performance of the model depending on the difference in the distributions and the generalization capability of the model. It remains interesting to study drifts in the $P(X)$ distribution as it is observable at prediction time and is model-agnostic. As an example, the extension of the ML system usage to a new type of client can cause this kind of drift.
- $P_t(y|X) \neq P_{t+1}(y|X)$ while $P_t(X) = P_{t+1}(X)$, that is the distribution of input $P_t(X)$ remains unchanged, but the true decision boundary updates. Therefore, the decision boundary learned by the model is outdated and this causes a drop in accuracy in the region where the boundary has changed. For example, the changing economic context in which a financial system evolves can cause this type of drift.

– $P_t(X) \neq P_{t+1}(X)$ and $P_t(y|X) \neq P_{t+1}(y|X)$, that is, both the input data distribution and the decision boundary changes. In many real-world applications, this type of drift occurs, and we observe a drop in model accuracy.

2.2 Drift Detectors

A drift detector is a method that observes a stream of data over time and determines for every new data point if the current distribution of the data has changed compared to a reference data set. We survey the literature for available drift detectors and identify three types of detectors: data-based detectors [11, 19,23], error-based detectors [2,4,6,8,16,20], and predictive detectors [10,21]. We select drift detectors that are scalable to handle our datasets, which contain more than a million examples. Our second criterion is the availability of the implementation - or the implementation details - that allows us to reproduce the detectors presented in their respective paper. We identified three data-based, seven error-based, and two predictive-based detectors. The intuition behind each detector and their tunable parameters are given in Appendix A.

2.3 Domain Generalization

The domain generalization (DG) problem was first formally introduced by Blanchard et al. [5]. Unlike other related learning problems such as domain adaptation or transfer learning, DG considers the scenarios where target data is *inaccessible* during model learning. Hence, adaptation to distribution shift falls under the umbrella of domain generalization. [24] introduced a categorization of techniques commonly used to address the DG challenge, including domain alignment training, synthetic data augmentation, and self-supervised learning. Our work stems from the need of our industrial partner to improve the monitoring of the deployed models and to effectively trigger their well-established retraining procedures. While we believe that DG techniques could also improve the performance of the models against distribution shift, all these approaches are orthogonal to our investigations for an efficient retraining schedule under delay.

2.4 Delays in Time Series Evaluation

Masud et al. [15] study the problem of novel class detection while considering the true label delay constraints. They show how delaying the classification of incoming samples helps to detect new classes before the true label is available, hence providing a more accurate prediction. Poenaru-Olaru et al. [18] investigated recently the reliability of data and error-based drift detectors. However, the experimental protocol used does not consider label and deployment in production delays, which is the core of our study. In [17], Plasse et al. introduced a taxonomy to describe the labeling delay mechanism. They showed how delayed labels can be used to pre-update classifiers in real-world applications. However, the study didn't tackle validation delays and their impact on the model's deployment or the drift monitoring process. Žliobaité [25] analyzed the factors that allow concept drift detection before the label is available. However, no experiments are conducted on the impact of the finding on the performance of ML predictions, which is the aim of our novel protocol.

3 Problem

Without loss of generality, we consider a classification problem defined on a n dimensional feature space $\mathcal{X} \subseteq \mathbb{R}^n$ and a binary label space $\mathcal{Y} = \{0,1\}$. We assume that the samples come as a time series S where each sample $(x_i, y_i) \in \mathcal{X} \times \mathcal{Y}$ is indexed by a discrete time parameter t_i, such that t_i represents the time at which the input x_i reached the system. The *labeling delay* of x_i is the time between t_i and the moment x_i receives its true label y_i. For simplicity, we assume that this delay δ_l is constant across the inputs.

Let $h_{t_j} : \mathcal{X} \to \mathcal{Y}$ be the classification model trained at time t_j. Then h_{t_j} can only be trained on inputs $\{x_i\}$ such that $t_i + \delta_l \leq t_j$. Furthermore, the *deployment delay* of h_{t_j} is the time needed to deploy it in production. As before, we assume that this delay δ_d is constant. Thus, any model h_{t_j} can only make predictions on inputs that arrive in the system after it is deployed, that is, on inputs $\{x_k\}$ such that $t_j + \delta_d \leq t_k$.

We define a *retraining schedule* as an ordered sequence $sched = \{t_1 \ldots t_n\}$ that determines when a model should be re-trained. For example, periodic retraining uses a sequence in which all elements are exactly separated by a constant period p, that is, $t_{j+1} = t_j + p$. By contrast, drift detectors decide the schedule on the fly based on their statistical analysis of the data and the model. A retraining schedule determines a sequence of models $H = \{h_{t_1} \ldots h_{t_j} \ldots h_{t_n}\}$. Then, any input example x_i observed at time t_i is predicted by the latest available model: The predicted label for x_i is given by $\hat{y}_i = h_{t_*}(x_i)$ where $t_* = \max\{t_k \in sched \text{ s.t. } t_k + \delta_{prod} \leq t_i\}$.

The evaluation of retraining schedule is two-folds: *effectiveness* and *efficiency*. Effectiveness is measured as the negative of the prediction error made by the sequence of models H calculated by an arbitrary scoring function $score(\mathcal{Y}, \hat{Y})$, with $\hat{Y} = \{\hat{y}_i\}$. Efficiency measures the overall cost of model retraining. For simplicity, in our study we assume this cost to be constant across models and compute it as the number $n = |H|$ of retraining. This assumption matches the context of our partner, where the cost of deploying models in production largely surpasses the computational cost to retrain the model and the data labeling cost.

4 Methodology

We propose a novel protocol to thoroughly evaluate retraining scheduling techniques under realistic industrial constraints (labeling and deployment delays). These constraints have been overlooked by previous studies. We do so, moreover, while carefully and empirically considering alternative design decisions that affect model performance (incl. hyperparameter tuning and training window size).

Our protocol starts from an initial model m_0 trained on an initial training set $S_{train} =]d_0, d_{N_{train}}]$, which contains the first N_{train} example of the time series S. The protocol evaluates this model on the remaining samples of the time series $S_{test} = [d_{N_{train}}, d_{|S|}[$ using an arbitrary score function, which takes as input the prediction of the model and the true label.

4.1 Model Hyperparameter Tuning

The initial part of our protocol investigates whether hyperparameter tuning can improve the baseline model and if repeating this tuning at each retraining improves the model's effectiveness.

We select the best hyperparameter tuning strategy across three strategies. "No tuning" uses the hyperparameters provided by our industrial partner. "Initial tuning" involves training the baseline model from scratch on the training set S_{train} with hyperparameter tuning, then keeping these hyperparameters fixed during evaluation. "Re-tuning" means tuning the hyperparameters each time the model is retrained, based on the data available at that time.

Hyperparameter tuning is performed using K-fold validation with time series splits and Bayesian search, aiming to maximize the average score function over the folds. Time series splits are preferred over random or stratified K-fold splits as they better evaluate the model's generalization to future samples, which is particularly useful for handling data drift and improving model robustness. We identify the **best tuning strategy** and reuse it for the rest of our protocol.

4.2 Training Window Size

We next need to use an appropriate window size is, i.e. how many of the most recent data model retraining should use. Training with the maximum amount of data does not always produce the best model, according to the dilemma between model plasticity (learning new information) and stability (retaining previous knowledge) [7,13].

We thus compare the effectiveness achieved by different window sizes that ranges from a fraction of the data up to all the available data. We select the best window size with periodic retraining schedule with different period ranging from a period as short as the label delay up to a period corresponding to a single retraining across the entire time series. We consider scenarios without and with delays. An evaluation without delays measures maximal potential performance improvements of alternative window sizes, whereas an evaluation with delays measures the improvement in a realistic context. Thanks to this step, we select the **best window size** and use it to evaluate retraining schedules.

4.3 Drift Detector Evaluation

We evaluate drift detectors and their parameters in a realistic scenario. We start from an initial model m_0 trained on the training set S_{train} with the best tuning strategy and window size (as explained before). Once a new sample $d_i = x_i, t_i, y_i$ arrives in the system, we use the initial model m_0 to predict a label $\hat{y}_i$ for x_i. We feed our drift detectors with the feature x_i, the time t_i, the label y_i and/or the prediction $\hat{y}_i$, depending on the detector type. If the detector does not detect a drift, we do nothing and wait for the next samples to arrive and be processed.

If we detect a drift, we retrain a model with the latest data $]d_{i-window}, d_i]$. In practice, we use a pre-trained model for evaluation such that d_i is rounded up. The newly trained model will become available for prediction after a delay $\delta = \delta_l + \delta_d$ with regard to t'_i. Between the detection of drift in t_i and the model being available in time $t_i + \delta$, we continue to predict the samples with m_0.

Whether we continue to detect drift or not depends on the type of drift detector. If the drift detector is data-based, we continue to detect drift. Indeed, these detectors do not rely on the model and can continue to trigger model retraining during this time window. On the other hand, error-based and predictive-based detectors observe model properties (e.g. error rate, uncertainty). Therefore, after detecting a drift, we wait for the new model and distribution to be available.

After the first drift has occurred, for the remainder of the sample $d_j, j > i$ we use the latest available model to make the prediction and follow the same process as in m_0. A model m_i is available at time t_{m_i} if and only if it has been trained with data older than $t_m - \delta_l - \delta_d$.

With these steps, we can evaluate the **drift detectors** effectiveness (according to the score function) and efficiency (number of retraining).

4.4 Comparing Drift Detectors and Periodic Retraining

To evaluate the periodic retraining strategy, we use the same protocol as for the drift detectors and instantiate a drift detector that is equivalent to periodic retraining. Note that this detector behaves like a data drift detector and can detect drift before the latest available model is used. To compare the efficiency (number of models) and effectiveness (ML metric) of periodic retraining schedules, we vary the retraining intervals to identify the optimal Pareto front.

For each detector, we tune its parameters using Bayesian search to minimize the number of models used and maximize the ML effectiveness on the test set. To avoid information leakage, we split the S_{train} set into K-fold of (s_{train}, s_{val}) using the time series split. For each fold, we evaluate the effectiveness and efficiency of the drift detector parameters using the protocol of Sect. 4.3. For each detector, we select the parameters on the Pareto front of efficiency and effectiveness.

We evaluate the effectiveness and efficiency of the kept parameters of all drift detectors on the complete dataset. We build the Pareto front of all the detectors' efficiency and effectiveness and compare them with the effectiveness and efficiencies we obtain with periodic retraining schedules.

These steps lead to the **best drift detector** and its best parameters.

5 Experiments

Below, we outline our empirical study protocol and evaluate each step's impact.

5.1 Dataset, Model and Metrics

We apply our empirical study to the transaction system of BGL BNP Paribas. The objective of the ML model is to classify a transaction as accepted or refused based on the recent transaction of a particular client. The dataset contains 1,093,587 labeled inputs from transactions that occurred over a 5.6-year period. The timestamps associated with inputs are precise for a single day.

The classifier previously developed by our partner is a random forest. To comply with the regulation, our partner must have the capacity to interpret the automated decision made by the model; and tree-based models are interpretable by design and, therefore, we use a random forest architecture like our partner. Due to the sensitivity of the system, we did not work with the real model in production but have created a baseline model with the support of our partner's instructions. Therefore, we built a random forest classifier with 100 estimators up to 8-level deep. We trained the baseline model with 400,000 samples, following our partner's recommendations. The minimum period for periodic retraining corresponds to the average label delay, which is 5,293. For simplicity, we round it down to 5,000 in our experiments.

We use Matthew's correlation coefficient (MCC) to score the prediction of our models which is well suited for unbalanced datasets. It is important to note that, in our partner's case, even small differences in MCC correspond to a significant business impact. For example, during our experiments, we noticed that a difference in 0.01 MCC corresponds to 3,000 transactions on average.

Figure 1 (blue curve) shows the performance of the baseline model over time, without retraining. This reveals that the model performance is not stable over time and tends to degrade after a certain point due to distribution drifts. On the first 20,000 samples, our model has an MCC of 0.5595. Later, the MCC score ranges between 0.5169 and 0.6099, which is a significant difference business-wise and alerts our partner. We observe a peak performance on the batch [740, 760[. After this peak, performance tends to degrade until an MCC of 0.5443 for the last batch of our evaluation. The orange curve shows the performance of the best model in our study; this shows the potential benefits that appropriate retraining can produce.

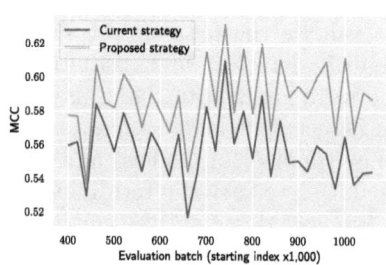

Fig. 1. Evolution of ML effectiveness (MCC) with time (in batches of 20,000 inputs). (Color figure online)

Additionally, we evaluate the LCLD dataset [12]. The task of the ML model is to classify loan requests as accepted or refused based on information provided by the client (e.g. purpose of the loan), publicly available data (e.g. credit score), and computed features (e.g. installment). The dataset contains 1,124,606 labeled inputs from transactions that occurred over a 5-year period.[1]

---

[1] Code available at https://github.com/serval-uni-lu/drift-robustness.

5.2 Results

Tuning the Hyperparameters of the Initial Model has a Positive Impact. We first investigate the benefits of re-tuning the model hyperparameters over time. We consider three different scenarios: 1) the baseline model of our partner is used throughout ("no tuning"); 2) we tune the model based on a time split[2] ("initial tuning"); and 3) the model is re-tuned each time it is retrained ("re-tuning"). Since we want to assess the potential of model tuning to get better models over time, we ignore labeling and deployment delays at this stage.

Table 1. Impact of training strategy on ML effectiveness.

Dataset	Model hyperparameters	Retraining period				
		20,000	50,000	100,000	200,000	400,000
BGL	No tuning	0.5678	0.5656	0.5648	0.5627	0.5616
	Initial tuning	**0.5887**	**0.5877**	**0.5864**	**0.5855**	**0.5842**
	Re-tuning	0.5882	0.5867	0.5862	0.5848	0.5837
LCLD	No tuning	0.2691	0.2688	0.2681	0.2678	0.2669
	Initial tuning	0.2742	0.2737	0.2725	**0.2723**	**0.2709**
	Re-tuning	**0.2751**	**0.2744**	**0.2734**	0.2711	0.2685

Table 1 shows that initial model tuning yields significant improvements but that re-tuning at each retraining is not necessary. For our partner model, initial tuning always has a higher MCC than no-tuning and re-tuning regardless of the retraining period. For LCLD, initial tuning is more effective for large retraining periods while re-tuning is more effective with frequent retraining. However, the MCC gains are always under 0.001 between initial and re-tuning. For our study, this means that we may proceed with a careful **initial tuning and skip re-tuning the model each time we retrain it**.

Training with all Available Data is Counterproductive. We evaluate the impact of the retraining window on the ML effectiveness when using the simple periodic retraining strategy. As before, we study this impact in the ideal scenario without delay to measure the maximum potential gains and in a realistic scenario with our partner delays.

Table 2 reveals that using 400k (respectively 200k) for our partner use case (respectively, LCLD) is the best retraining strategy on average. Breaking down the results for each period on our partner use case reveals that using the 400k most recent samples to retrain is always among the 2 best solutions without delay and the best solution with delays, independently of the retraining period.

---

[2] The baseline model of our partner is tuned on non-time-sensitive k-fold validation.

Table 2. Impact of window size on ML effectiveness. Average over periodic retraining with periods $p = \{5, 10, 20, 50, 100, 200, 400\} \times 10^3$.

	BGL		LCLD	
Delay Window size	No	Yes	No	Yes
50k	0.5737	0.5723	0.2726	0.2721
100k	0.5819	0.5804	0.2735	0.2729
200k	0.5867	0.5849	**0.2741**	**0.2732**
400k	**0.5875**	**0.5857**	0.2732	0.2726
All	0.5860	0.5841	0.2733	0.2730

We also observe that using all the available data is only the third best solution – for all periods except 50k – and is therefore not the optimal solution. Consequently, we empirically set the retraining window used in our experiments at 400k examples for our partner and 200k for LCLD (Table 3).

Periodic Retraining and Error-Based Detectors Together Offer a Flexible Compromise Between Effectiveness and Efficiency. We investigate the effectiveness and efficiency of drift detectors and periodic retraining schedules in scenarios with delays of BGL BNP Paribas, comparing their efficiency (number of retrainings) and effectiveness (MCC) in Fig. 2 and Table 4 In Fig. 2, each data point is a particular method setting, i.e. a scheduling method (periodic retraining or drift detector) with given parameter values. All scheduling methods (periodic and detectors) appear on the Pareto front, including the no-retraining strategy due to its inherent efficiency. We excluded drift detector schedules equivalent to "no retraining" or "always retraining". Figure 3a indicates that for a retraining budget above 22, error-based detectors are as effective as periodic retraining with fewer retrainings. For instance, the HDDM-W detector performs better with 32 retrainings compared to the most effective periodic strategy requiring 136 retrainings. Periodic retraining is most effective for up to seven retrainings, with no clear advantage for either method between seven and 22 retrainings. Similarly, for LCLD, data-drift detectors and periodic retraining are most effective for fewer retrainings, and error-based detectors for higher numbers.

In the second column of Table 4, for each method, the right number shows the number of parameter settings that are Pareto-optimal *within* this method (i.e., the best settings of this particular method); the left number shows the number of these settings that are Pareto-optimal *across* all methods. The statistical test and PCA-CD detectors fail to reach the Pareto front for Partner, while the divergence detector succeeds. Error-based detectors DDM, EDDM, and Page-Hinkley (PE) also fail for both datasets, but predictive-based detectors reach the Pareto front.

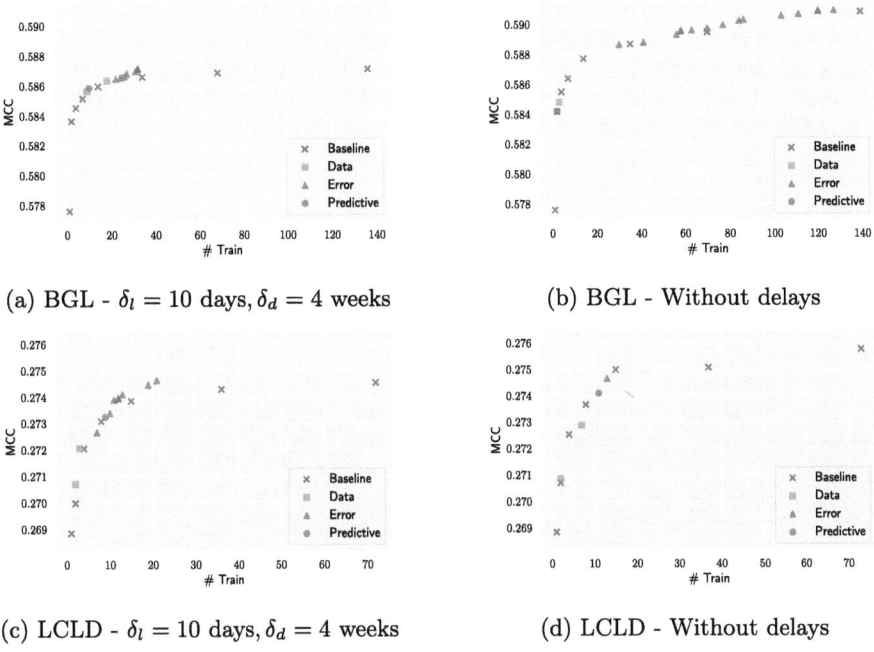

Fig. 2. Pareto front of drift detectors, no retraining VS periodic retraining.

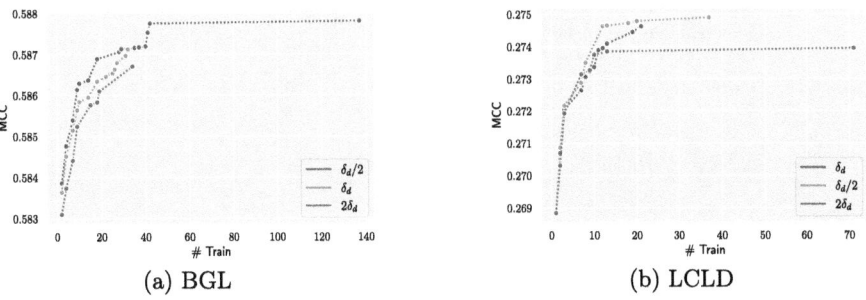

Fig. 3. Pareto front of retraining schedule with different deployment delays.

5.3 Generalization Study

Not Considering Delay Overestimates the Effectiveness/Efficiency Trade-Off of Retraining Schedules. To emphasize the importance of considering the delay in the evaluation of retraining schedules, we compare the drift detector and periodic retraining schedules with and without delay. We find that not considering delay overestimates the effectiveness/efficiency trade-off of retraining schedules. Indeed, comparing Fig. 3a (respectively 3c) with Fig. 3b (respectively 3d), we observe that the Pareto front lies higher when there is no delay. For example, on our partner's case the most effective strategy without

Table 3. For each schedule, the number of parameter settings (A) / (B). (A) are the settings on the efficiency/effectiveness Pareto front across all methods. (B) are the settings on the Pareto front local to the method. For varying delays, we report the number of Parameters on the Pareto front with delay δ_d that remains on the Pareto front for delays $\delta_d/2$ and $2\delta_d$. $\delta_l = 10$ days, $\delta_d = 4$ weeks.

Type	Schedule	BGL				LCLD			
		No	δ_l δ_d	δ_l $\delta_d/2$	δ_l $2\delta_d$	No	δ_l δ_d	δ_l $\delta_d/2$	δ_l $2\delta_d$
Baseline	No detection	1 / 1	1 / 1	1	1	1 / 1	1 / 1	1	1
	Periodic	4 / 7	4 / 7	4	3	5 / 7	1 / 7	0	1
Data-based detector	Statistical test	0 / 25	0 / 25	0	0	0 / 9	0 / 9	0	0
	Divergence	1 / 4	2 / 6	2	2	1 / 4	1 / 5	2	1
	PCA-CD	0 / 3	0 / 4	0	0	1 / 2	1 / 2	1	1
Error-based detector	ADWIN (CE)	3 / 9	1 / 4	0	0	1 / 3	3 / 6	2	1
	ADWIN (PE)	1 / 9	1 / 3	0	0	0 / 5	0 / 5	0	0
	DDM	0 / 4	0 / 3	0	0	0 / 5	0 / 2	0	0
	EDDM	0 / 3	0 / 4	0	0	0 / 6	0 / 6	0	0
	HDDM-A	5 / 6	10 / 12	0	10	0 / 19	0 / 4	1	1
	HDDM-W	2 / 2	17 / 17	1	16	0 / 1	0 / 17	0	0
	KSWIN (CE)	1 / 8	0 / 7	0	0	0 / 7	2 / 5	3	0
	KSWIN (PE)	1 / 3	1 / 3	0	0	0 / 4	2 / 4	0	0
	Page-Hinkley (CE)	2 / 8	1 / 3	0	0	0 / 1	0 / 2	0	0
	Page-Hinkley (PE)	0 / 3	0 / 2	0	0	0 / 1	0 / 1	0	1
Predictive-based detector	Uncertainty	1 / 14	1 / 11	1	0	0 / 3	0 / 5	0	0
	Aries ADWIN	0 / 4	1 / 4	0	0	1 / 5	1 / 3	0	1

delays has an MCC of 0.5909 for Page-Hinkley (CE), while with delays, the most effective strategy has an MCC of 0.5871 for HDDM-W.

The Relative Ranking Between the Methods also Changes. Table 4 compares the number of method settings (for each method) that are on the Pareto front, in the cases without delays (left column) and with delays (right column). We see that the scheduling method settings on the Pareto front are different in the two cases. For instance, for our partner use case, KSWIN (CE) had one setting on the Pareto front without delay, but this setting disappears from the front when there is a delay. Hence, the optimal drift detection method settings without delay do not remain optimal if delays occur.

Change in Deployment Delay has an Opposite Effect on the Effectiveness/Efficiency Pareto Front. We study the impact of varying delays on the effectiveness and efficiency of retraining scheduling methods by simulating scenarios where the deployment delay is increased or decreased after the methods are tuned and running. We tune the detectors based on the previously used delays δ_l and δ_d and then evaluate them when δ_d is halved or doubled. Figure 3 compares the Pareto fronts in the cases where δ_d is halves, unchanged and doubled. We observe that the $\delta_d/2$ front dominates the δ_d front, which itself dominates the $2\delta_d$ one. This indicates that a reduction in deployment delay (compared to the delay considered when tuning the drift detectors) yields improvement, whereas an augmented delay incurs a loss in the effectiveness/efficiency trade-off.

Changes in Deployment Delays Disrupt the Relative Ranking of the Retraining Scheduling Methods. Table 4 shows the number of schedules (method settings) that are on the Pareto front for the original delay δ_d (second column (A)) and how many of these exact schedules remain on the front for delays $\delta/2$ (third column) and 2δ (fourth column). A retraining schedule generalizes if it remains on the Pareto front in spite of the delay change. For our partner use case, we observe that only three methods generalize to the halved and doubled delays: periodic retraining, HDDM-W, and divergence. The uncertainty detector only generalizes to a reduction of the delays. All other methods do not generalize. For HDDM-W, the parameters that generalize when augmenting the delay are different than the ones when reducing the delay. On the LCLD dataset, only Divergence and ADWIN generalize to the halved and doubled delays.

5.4 Generalization Beyond Financial Domain

Electricity Dataset. Electricity [9] is a widely used dataset in the distribution shift on tabular data literature as shown in [14]. The classification task is to determine at any point in time whether the electricity price is going up or down. The six features include the day of the week, the current price, the electricity demand, as well as the price, demand, and transfer of the adjacent geographical region. This dataset is smaller than the financial datasets, spans over a shorter period, and is drawn from another domain. Electricity contains 45,312 labeled inputs that occur over 943 days with exactly one input recorded every 30 min. The dataset is precise to 30 min. Hence, the labeling and deployment delays may differ in a real-world system. The minimum period of retraining corresponds to the average label delay, which is 480 samples. We trained the baseline model with one year of data. We round up to the next multiple of 480 to facilitate model reuse during the experiments and obtain 17760 inputs.

Results. We started with the same delays as for financial datasets ($\delta_l = 10$ days, $\delta_d = 4$ weeks). With these settings, we observe in Fig. 5a that none of the schedules, including retraining every 10 days, can outperform the baseline. Not retraining and keeping the original model is the most effective strategy for such delays. This confirms the importance of considering the delay in the evaluation of retraining schedules. We also consider a scenario with shorter delays. We use

Table 4. For each schedule, the number of parameter settings (A) / (B). (A) are the settings on the efficiency/effectiveness Pareto front across all methods. (B) are the settings on the Pareto front local to the method. For varying delays, we report the number of Parameters on the Pareto front with delay δ_d that remains on the Pareto front for delays $\delta_d/2$ and $2\delta_d$. $\delta_l = 1$ day, $\delta_d = 10$ days.

Type	Detector	Delay			
		No	$\delta_l = 1$ day $\delta_d = 9$ days	$\delta_d/2$	$2\delta_d$
Baseline	No detection	1 / 1	1 / 1	1	1
	Periodic	1 / 7	1 / 7	2	1
Data-based detector	Statistical test	2 / 9	6 / 7	4	7
	Divergence	1 / 4	0 / 5	1	0
	PCA-CD	0 / 0	0 / 0	0	0
Error-based detector	ADWIN (CE)	0 / 10	3 / 4	0	2
	ADWIN (PE)	0 / 4	3 / 4	0	0
	DDM	1 / 3	0 / 3	0	0
	EDDM	0 / 4	0 / 4	0	0
	HDDM-A	0 / 23	7 / 7	1	0
	HDDM-W	1 / 2	0 / 2	1	2
	KSWIN (CE)	0 / 5	3 / 5	0	0
	KSWIN (PE)	0 / 9	1 / 2	1	0
	Page-Hinkley (CE)	0 / 2	0 / 2	2	0
	Page-Hinkley (PE)	1 / 1	0 / 1	1	1
Predictive-based detector	Uncertainty	1 / 4	4 / 4	3	4
	Aries ADWIN	1 / 4	1 / 3	0	0

a total delay of 10 days corresponding to our previous label delay. We assume that at any given time the electricity price of the previous day is available. Hence, we split this total delay into 1 day for labeling and 9 days for production. With these settings, we observe that error-based detectors are the best trade-off between accuracy and efficiency for 12 or more retrains. Below 12 retrains, not retraining remains the best strategy.

When comparing the occurrences on the Pareto front of each schedule strategy between the scenario with and without delays on Table 4 for Electricity, we confirm that the optimal schedule varies. KSWIN, HDDM-A, and ADWIN become relevant for Electricity. Similarly, changing the delay has an impact on which schedule strategy remains optimal. ADWIN and KSWIN (CE) are no longer on the Pareto front when halving the delay for Electricity and HDDM-A is no longer on the front when we double the delay.

Similarly to finance use cases, doubling or halving the delay has a significant impact on the Pareto-front of drift detectors. In Fig. 5, the MCC decreases by more than 5% between scenarios $\delta_d/2$ and $2\delta_d$ Fig. 4.

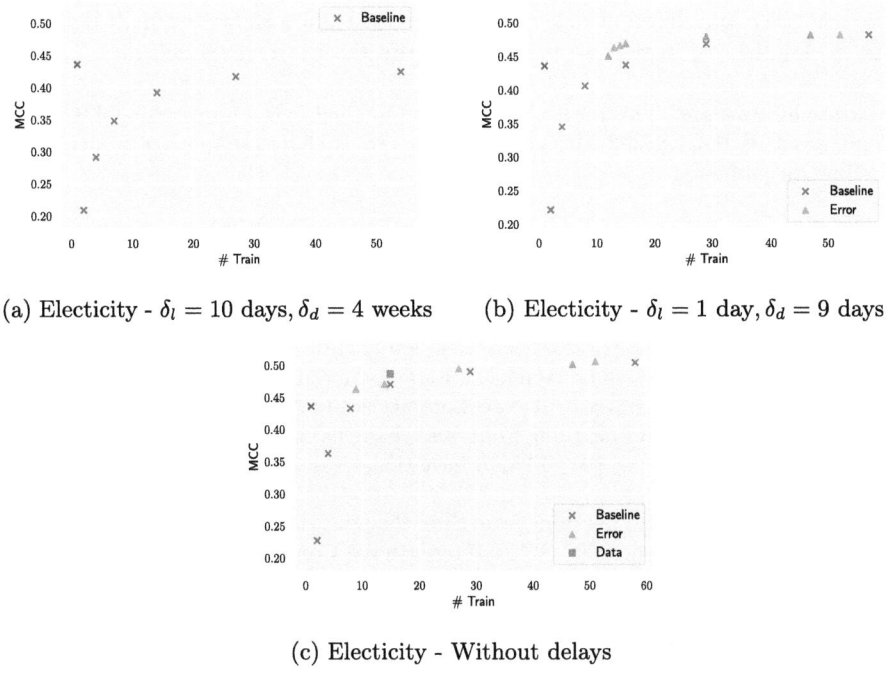

Fig. 4. Pareto front of drift detectors, no retraining and periodic retraining schedules on Electricity dataset.

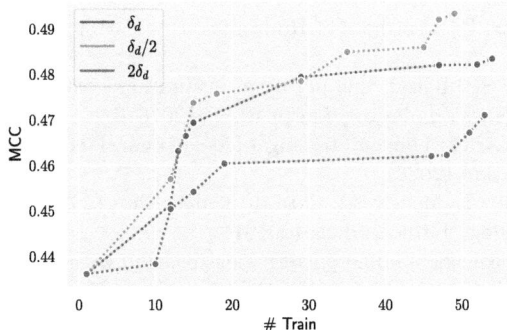

Fig. 5. Pareto front of retraining schedule with same parameters and different deployment delays.

6 Conclusion

In this paper, we studied the performance of ML models with drift detector triggered retrain in the presence of delays. We considered industrial use cases where the label arrives ten days after the prediction and the model goes through a four-week validation phase before deployment. We evaluated 15 detectors in two use cases and different delay scenarios. Our results show that drift detectors and scheduling strategies are particularly affected by realistic delays.

Acknowledgments. This project is supported by the Luxembourg National Research Fund, grant BRIDGES/2022/IS/17437536. This research was supported by BGL BNP Paribas Luxembourg.

References

1. Amershi, S., et al.: Software engineering for machine learning: a case study. In: Proceedings - ICSE-SEIP 2019, pp. 291–300 (2019)
2. Baena-García, M., Campo-Ávila, J., Fidalgo-Merino, R., Bifet, A., Gavald, R., Morales-Bueno, R.: Early drift detection method (2006)
3. Bifet, A., Gavaldà, R.: Learning from time-changing data with adaptive windowing. In: Proceedings of the 7th SIAM International Conference on Data Mining, pp. 443–448 (2007)
4. Bifet, A., Gavaldà, R.: Learning from time-changing data with adaptive windowing. In: Proceedings of the 2007 SIAM International Conference on Data Mining, pp. 443–448. Society for Industrial and Applied Mathematics (2007)
5. Blanchard, G., Lee, G., Scott, C.: Generalizing from several related classification tasks to a new unlabeled sample. In: NeurIPS (2011)
6. Frias-Blanco, I., Campo-Avila, J.D., Ramos-Jimenez, G., Morales-Bueno, R., Ortiz-Diaz, A., Caballero-Mota, Y.: Online and non-parametric drift detection methods based on Hoeffding's bounds. IEEE Trans. Knowl. Data Eng. **27**(3), 810–823 (2015)
7. Gama, A., Bifet, A., Barcelona, R.: A survey on concept drift adaptation. ACM Comput. Surv **46** (2014)
8. Gama, J., Medas, P., Castillo, G., Rodrigues, P.: Learning with drift detection. **8**, 286–295 (2004)
9. Harries, M.B.: SPLICE-2 comparative evaluation: electricity pricing (1999). https://api.semanticscholar.org/CorpusID:151207670
10. Hu, Q., et al.: Aries: efficient testing of deep neural networks via labeling-free accuracy estimation (2023)
11. Evidently AI Inc.: Evidently AI: data drift algorithm (2021)
12. Kaggle: All lending club loan data (2019)
13. Lim, C.P., Harrison, R.: Online pattern classification with multiple neural network systems: an experimental study. IEEE Trans. Syst. Man Cybern. Part C Appl. Rev. **33**(2), 235–247 (2003)
14. Lu, J., Liu, A., Dong, F., Gu, F., Gama, J., Zhang, G.: Learning under concept drift: a review. IEEE Trans. Knowl. Data Eng. **31**(12), 2346–2363 (2019)
15. Masud, M., Gao, J., Khan, L., Han, J., Thuraisingham, B.M.: Classification and novel class detection in concept-drifting data streams under time constraints. IEEE Trans. Knowl. Data Eng. **23**(6), 859–874 (2010)
16. Page, E.S.: Continuous inspection schemes. Biometrika **41**(1/2), 100–115 (1954)
17. Plasse, J., Adams, N.: Handling delayed labels in temporally evolving data streams. In: 2016 IEEE International Conference on Big Data (Big Data), pp. 2416–2424. IEEE (2016)
18. Poenaru-Olaru, L., Cruz, L., van Deursen, A., Rellermeyer, J.S.: Are concept drift detectors reliable alarming systems? – A comparative study (2022)

19. Qahtan, A.A., Alharbi, B., Wang, S., Zhang, X.: A PCA-based change detection framework for multidimensional data streams. In: Proceedings of the 21th ACM SIGKDD International Conference on Knowledge Discovery and Data Mining, pp. 935–944. ACM, Sydney NSW Australia (2015)
20. Raab, C., Heusinger, M., Schleif, F.M.: Reactive soft prototype computing for concept drift streams. Neurocomputing **416** (2020)
21. Shaker, M.H., Hüllermeier, E.: Aleatoric and epistemic uncertainty with random forests (2020)
22. Storkey, A., et al.: When training and test sets are different: characterizing learning transfer. Dataset Shift Mach. Learn. **30**(3–28), 6 (2009)
23. Van Looveren, A., et al.: Alibi detect: algorithms for outlier, adversarial and drift detection (2019)
24. Zhou, K., Liu, Z., Qiao, Y., Xiang, T., Loy, C.C.: Domain generalization: a survey. IEEE Trans. Pattern Anal. Mach. Intell., 1-20 (2022). https://doi.org/10.1109/tpami.2022.3195549
25. Žliobaitė, I.: Change with delayed labeling: when is it detectable? In: 2010 IEEE International Conference on Data Mining Workshops, pp. 843–850 (2010)

Understanding Knowledge Drift in LLMs Through Misinformation

Alina Fastowski(✉) and Gjergji Kasneci

Technical University of Munich, Munich, Germany
`alina.fastowski@tum.de`

Abstract. Large Language Models (LLMs) have revolutionized numerous applications, making them an integral part of our digital ecosystem. However, their reliability becomes critical, especially when these models are exposed to misinformation. This paper primarily analyzes the susceptibility of state-of-the-art LLMs to factual inaccuracies when they encounter false information in a Q&A scenario, an issue that can lead to a phenomenon we refer to as *knowledge drift*, which significantly undermines the trustworthiness of these models. We evaluate the factuality and the uncertainty of the models' responses relying on Entropy, Perplexity, and Token Probability metrics. Our experiments reveal that an LLM's uncertainty can increase up to 56.6% when the question is answered incorrectly due to the exposure to false information. At the same time, repeated exposure to the same false information can decrease the models' uncertainty again (-52.8% w.r.t. the answers on the untainted prompts), potentially manipulating the underlying model's beliefs and introducing a drift from its original knowledge. These findings provide insights into LLMs' robustness and vulnerability to adversarial inputs, paving the way for developing more reliable LLM applications across various domains. The code is available at https://github.com/afastowski/knowledge_drift.

Keywords: Uncertainty · Knowledge Drift · Large Language Models

1 Introduction

The rapid advancement in natural language processing (NLP) has seen significant strides with the development of large-scale language models, such as the GPT family. These models have demonstrated remarkable capabilities in various tasks, including text generation, translation, and question-answering (QA). However, despite their impressive performance, critical challenges remain in understanding and improving these models' reliability and robustness, especially regarding their factual knowledge and uncertainty estimation.

Understanding the reliability of language models is crucial, particularly in applications where the consequences of incorrect or uncertain answers can be significant. For instance, in fields like healthcare, law, and education, the ability to trust the outputs of a language model is paramount. One key aspect of this reliability is the model's ability to handle false or misleading information. By evaluating how language models respond to false information, we can gain insights into their internal knowledge structures and the robustness of their factual accuracy. With our work, we are assessing what we refer to as the *knowledge drift* of these models, by which we refer to any changes in their internal knowledge and beliefs. In our case, we are interested in models' knowledge drift as a result of manipulative user interactions.

Specifically, we examine the impact of knowledge drift on model answer uncertainty by studying the effect of false information presented to it with the prompt. By leveraging the TriviaQA dataset [10], we aim to analyze how the model's performance varies with different types of misleading information. Such insights can help us identify potential vulnerabilities in its knowledge processing.

Contributions. In this paper, we contribute to this body of research by evaluating LLMs' – i.e., GPT-4o, GPT-3.5, LLaMA-2-13B, and Mistral-7B – responses to false information in a QA task setting. By analyzing the models' uncertainty under various metrics and answer accuracy under different conditions, we aim to shed light on the robustness of their knowledge obtained during pre-training. This paper provides three main contributions to this topic:

1. *Impact of False Information on Uncertainty.* We investigate how introducing false information into question prompts affects LLMs' performance and uncertainty estimation in a QA setting. Our analysis reveals that while false information initially increases the model's uncertainty, repeated exposure can lead to decreased uncertainty, indicating successful manipulation and drift of the model away from its original, correct beliefs.
2. *Effect of Random Information.* We demonstrate that random, unrelated information results in the highest levels of model uncertainty, suggesting that the model experiences greater confusion with irrelevant data than with targeted false information. This finding underscores the importance of context relevance in understanding the model's responses.
3. *Insights into Model Vulnerabilities and Robustness.* Our study provides critical insights into the vulnerabilities of LLMs to adversarial inputs. By highlighting the limitations of current uncertainty estimation methods in adversarial attack detection, our work contributes to efforts aimed at enhancing the robustness and trustworthiness of language models for practical applications.

2 Related Work

Language Model Architectures and Capabilities. Understanding LLM knowledge drift necessitates grounding in the foundational works on language models. Pioneering efforts such as GPT [18], BERT [6], or T5 [19] remain fundamental to contemporary LLMs. This groundwork facilitated the recent advancements in LLM capabilities, exemplified by Brown et al.'s [2] exploration of few-shot learning and the development of powerful models like GPT-4 [1], or PaLM [4] with its

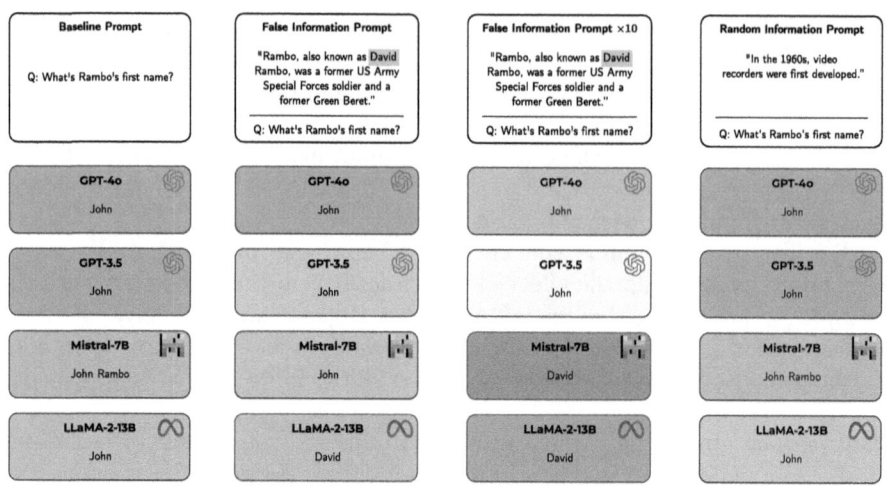

Fig. 1. Answers produced by state-of-the-art LLMs on *"What's Rambo's first name?"* with no perturbation (col. 1), with false information injection (cols. 2 & 3), and with random information injection (col. 4). Green boxes indicate correct answers; red are incorrect. The transparency of the boxes indicates the uncertainty of the model: i.e., the lighter, the more uncertain. Notice how injecting the same false information multiple times makes LLMs more uncertain (see GPT-3.5) and can even shift their original correct answer to a wrong one (see Mistral and LLaMA).

pathway architecture. Furthermore, the emergence of efficient and open-source models like LLaMA [23] and its successors highlight the ongoing progress in LLM accessibility and development.

Uncertainty in LLMs. While prior work has explored uncertainty quantification in NLP tasks like calibration of classifiers and text regressors ([5,7,9,24]), these approaches often rely on techniques directly transferable from other domains (e.g., Monte Carlo dropout, Deep Ensembles). However, as highlighted by [12], generative tasks in NLP present unique challenges due to semantic equivalence. For instance, Jiang et al. [9] demonstrate a weak correlation between answer confidence (log-likelihood) and correctness in generative question answering.

Recent efforts have tackled uncertainty or calibration in Natural Language Generation (NLG) by prompting models to assess their outputs or fine-tuning models to predict uncertainty ([11,13,14]). While these methods can be effective, they often require additional training data and supervision, leading to challenges in reproducibility, cost, and sensitivity to distribution shifts (e.g., hardware limitations preventing implementation as in [11]).

The challenges associated with uncertainty estimation in NLG mirror those in automatic NLG evaluation. For example, Ott et al. [15] highlight performance limitations in machine translation due to multiple valid translations for a single source sentence. Similarly, Sai et al. [22] discuss the potential of paraphrase detection for NLG evaluation, which may offer insights applicable to uncertainty estimation tasks.

LLM Factual Knowledge and Calibration. Understanding LLM knowledge drift requires examining research on factual knowledge capabilities and calibration in question answering. Roberts et al. [21] investigate the capacity of LLMs to retain factual knowledge implicitly learned during pre-training, proposing methods to assess this internal knowledge store. Youseff et al. [25] take a broader view, surveying various techniques for probing factual knowledge within pre-trained LLMs. Building on this, Petroni et al. [17] explore the potential of LLMs as factual knowledge bases, analyzing their ability to serve as information sources. Jiang et al. [9] address the crucial calibration issue in LLM question answering. Their work examines how well LLM confidence scores align with answer correctness, a critical aspect of reliable knowledge extraction.

Interestingly, when LLMs are asked to answer a particular question, they often produce agreeable responses that align with input biases, even if factually incorrect, also called *sycophancy* in the literature [8,16,20]. These findings underscore the need for models to critically assess and counter false inputs to maintain reliability and factual accuracy. Aligned with these works, we prompt SoTA models with false information and observe that they still exhibit seemingly sycophantic behaviour w.r.t. the question and engender ungrounded responses, leading to performance drops.

3 Experiments

In this work, we evaluate the factual knowledge of recent large language models and their associated uncertainty levels in a Question Answering (QA) task setting. The objective is understanding how false information embedded in the question prompts influences the models' performance and uncertainty metrics. We expect that the more false information is fed to these language models, the more certain they will become about it while giving up on accuracy since they begin generating false information.

3.1 Experimental Setup

Dataset and Models. We have two requirements for choosing suitable LLMs for our experiments: 1) performing reasonably well on closed-book question-answering without additional fine-tuning, and 2) providing access to the log probabilities of the generated tokens. Hence, we experiment with GPT-4o, GPT-3.5, Mistral-7B, and LLaMA-2-13B[1].

---

[1] Specifically, we use the following checkpoints: gpt-3.5-turbo-0125 and gpt-4o (accessed via the OpenAI API), Llama-2-13b-chat-hf and Mistral-7B-Instruct-v0.3 (accessed via huggingface).

Table 1. Performances of each LLM on 1000 samples of the TriviaQA dataset. Note that the number of parameters of GPT-4o has not been disclosed yet.

	Accuracy	#Parameters
GPT-4o	0.790	NA
GPT-3.5	0.721	1.75×10^{11}
Mistral-7B	0.502	7×10^9
LLaMA-2-13B	0.428	1.3×10^{10}

To assess the first requirement, we test the models on 1000 samples from the TriviaQA dataset [10] by prompting them with the given question. TriviaQA is a reading comprehension dataset consisting of question-answer-context triplets, where the contexts are text snippets containing the answer. In our setting, we ignore the contexts and use only the question-and-answer pairs. We also ask the model to respond only with the exact answer to avoid verbosity in the model's answers and ease the comparison with the correct answers.

Performance Evaluation. Our experiments are designed to assess two primary factors: i.e., 1) the correctness of the answers provided by the model, and 2) the uncertainty scores associated with the generated tokens.

We begin by identifying questions from the TriviaQA dataset that the LLM can answer correctly in a closed-book setting – i.e., the model answers without additional context or external information. This process allows us to focus only on the model's correct knowledge since we will later try to manipulate it. To assess the correctness of a response, we check if the true answer provided by the dataset is part of the model-generated answer. For example, the model might produce *"Chicago, Illinois"* as its answer, when the ground truth is *"Chicago"*. In this example, our process would check if *"Chicago"* is part of *"Chicago, Illinois"*, correctly marking the model answer as true (eventhough more verbose). Table 1 shows the accuracy of each model when prompted <u>once</u> with the questions from the dataset.

We intentionally chose LLMs of varying levels of QA performance, as this difference may lead to additional interesting insights about uncertainty developments. For the subsequent experiments, only the samples answered correctly by the model will be used for their respective evaluations.

Uncertainty Metrics. Given an input sequence x and parameters θ, an autoregressive language model generates an output sequence $y = [y_1, ..., y_T]$ where T is the length of the sequence. To quantify the model's uncertainty, we rely on *entropy* (1) and *perplexity* (2), as previously introduced by [3]. For calculating the entropy of each token, we take into account the top $i = 10$ probable tokens at each token position t. Lastly, as a more intuitively interpretable metric, we report the *probability* (3) of the generated tokens, averaged over all answer tokens. While the use of multiple metrics ensures the robustness in our measurements, they also capture slightly different dimensions: entropy focuses more on a token-level uncertainty, since we measure over multiple token options at each position, whereas perplexity and probability operate on more of a sentence level, simply averaging over all top-1-choice tokens in the generated sequence.

$$H(y \mid x, \theta) = -\frac{1}{T} \sum_t \sum_i p(y_{t_i} \mid y_{<t_i}, x) \log p(y_{t_i} \mid y_{<t_i}, x) \tag{1}$$

$$PPL(y \mid x, \theta) = \exp(-\frac{1}{T} \sum_t \log p(y_t \mid y_{<t}, x)) \tag{2}$$

$$TP(y \mid x, \theta) = \frac{1}{T} \sum_t \exp(\log p(y_t \mid y_{<t}, x)) \tag{3}$$

Baseline and Information Injection. We establish a baseline by evaluating the model's answers to the identified questions and their corresponding uncertainty scores. This baseline represents the model's performance and certainty without any misleading information. We then introduce false information into the question prompts and measure its impact on the model's answers and uncertainty scores. Notice that we do not provide the model with the correct answer or possible options for the posed question. Therefore, for visualization purposes, the following prompts contain a straight line that separates the actual prompt from the ground truth. Additionally, the words before the colon are not included – we use them to give context to the reader. We rely on two types of prompts:

- *False Info Prompt* (FIP): The prompt includes false information related to the question. For example:

> ✗ False Information: "Alfred Hitchcock directed 2001: A Space Odyssey."
> Question: "Who directed 2001: A Space Odyssey?"
> ──────────────────
> ✓ Correct Answer: "Stanley Kubrick"

- *Random Info Prompt* (RIP): The prompt includes random, unrelated information. For example:

> ✗ False Information: "Alfred Hitchcock directed 2001: A Space Odyssey."
> Question: "Who directed 2001: A Space Odyssey?"
> ──────────────────
> ✓ Correct Answer: "Stanley Kubrick"

We extend the above prompts with two different variants of instructions:

> **Prompt V1:**
> ⋮   ⋮   ⋮   ⋮
> Respond with the exact answer only.

> **Prompt V2:**
> ⋮ ⋮ ⋮ ⋮
> Respond with the <u>true</u>, exact answer only.

While the first prompt simply asks the model to answer the question, the second emphasizes choosing what the model believes to be the *factually correct* answer. This way, regardless of the injected false information in the overall prompt, we aim to reduce possible sycophantic behavior. Notice that the ⋮ represent either a FIP or RIP described above, followed by the question. In the above prompt instructions, *exact* aims to limit the answer's verbosity, while *true and exact* additionally emphasizes the truthfulness we wish to see from the LLM.

3.2 Results and Discussion

Evaluating Model Uncertainty and Knowledge Retention with Varying Prompt Integrity. Tables 2 and 3 show the changes in uncertainty scores according to Prompt V1 and V2, respectively. We report averages (± standard errors) over 10 runs on the questions each model answered correctly. So, for example, if GPT-4o answered correctly 79% of the time, we test its uncertainty on this data subset. We are aware that the correct answers two models produce[2] might be related to different questions, which, in turn, could result in unfair comparisons. Nevertheless, we argue that testing the model's uncertainty on inherently incorrect answers is insignificant since it does not provide any added value to studying the knowledge drift of LLMs. Therefore, we aim to verify how the correct knowledge shifts when prompts are tainted with false information.

It is interesting to note that we did not expect the models to produce incorrect answers on the baseline (B) (see Tables 2 and 3) since the prompt has not changed from the one used to identify the correct knowledge in Table 1. Still, when prompted multiple times, the models might engender wrong answers – we noticed a 1-2% drop of accuracy[3] w.r.t. what was reported in Table 1. However, as expected, all models have higher uncertainty scores on the incorrect answers for both prompt types on B. All the reported metrics reflect this, i.e., higher entropy, higher perplexity, and lower token probability indicate higher uncertainty.

These results are also consistent with FIP and RIP across the board, suggesting that false/random information has the same effect regarding correct vs. incorrect answers. Notice how, even when we inject false/random information into the prompts, the uncertainty levels on the correct answers remain similar (e.g., see GPT-3.5 B vs. FIP/RIP), suggesting that the correct model's knowledge remains unscathed. Contrarily, we notice a drop in the latter if we look at the uncertainty difference between B and FIP, e.g., in terms of entropy. We

---
[2] For instance, Mistral's 42.8% vs. GPT-3.5's 72.1%.
[3] For LLaMA, we observed a drop of 17%.

Table 2. Prompt V1 : Changes in uncertainty scores of the generated answers. The baseline represents the uncertainty when prompted with the question and giving the correct answer. – depicts missing samples to calculate metrics on.

		Correct Answers			Incorrect Answers		
		H ↓	PPL ↓	TP ↑	H ↓	PPL ↓	TP ↑
GPT-4o	B	$0.10^{\pm.004}$	$1.06^{\pm.003}$	$0.96^{\pm.002}$	$0.53^{\pm.102}$	$1.40^{\pm.096}$	$0.78^{\pm.041}$
	FIP	$0.12^{\pm.005}$	$1.07^{\pm.003}$	$0.95^{\pm.002}$	$0.23^{\pm.027}$	$1.17^{\pm.036}$	$0.90^{\pm.013}$
	RIP	$0.13^{\pm.005}$	$1.08^{\pm.003}$	$0.95^{\pm.002}$	$0.37^{\pm.044}$	$1.28^{\pm.056}$	$0.85^{\pm.020}$
GPT-3.5	B	$0.16^{\pm.007}$	$1.08^{\pm.005}$	$0.94^{\pm.003}$	$0.39^{\pm.073}$	$1.32^{\pm.088}$	$0.82^{\pm.038}$
	FIP	$0.16^{\pm.007}$	$1.08^{\pm.004}$	$0.94^{\pm.003}$	$0.27^{\pm.020}$	$1.16^{\pm.024}$	$0.90^{\pm.009}$
	RIP	$0.17^{\pm.007}$	$1.09^{\pm.005}$	$0.93^{\pm.003}$	$0.57^{\pm.045}$	$1.49^{\pm.070}$	$0.77^{\pm.020}$
Mistral-7B	B	$0.21^{\pm.009}$	$1.12^{\pm.007}$	$0.92^{\pm.004}$	–	–	–
	FIP	$0.21^{\pm.013}$	$1.10^{\pm.008}$	$0.93^{\pm.005}$	$0.23^{\pm.017}$	$1.13^{\pm.013}$	$0.92^{\pm.007}$
	RIP	$0.22^{\pm.011}$	$1.12^{\pm.008}$	$0.92^{\pm.004}$	$0.49^{\pm.036}$	$1.32^{\pm.036}$	$0.82^{\pm.015}$
LLaMA-2-13B	B	$0.18^{\pm.006}$	$1.09^{\pm.004}$	$0.94^{\pm.003}$	$0.30^{\pm.016}$	$1.17^{\pm.015}$	$0.89^{\pm.007}$
	FIP	$0.18^{\pm.008}$	$1.09^{\pm.005}$	$0.94^{\pm.004}$	$0.19^{\pm.007}$	$1.10^{\pm.005}$	$0.93^{\pm.003}$
	RIP	$0.19^{\pm.006}$	$1.09^{\pm.004}$	$0.93^{\pm.003}$	$0.33^{\pm.014}$	$1.20^{\pm.013}$	$0.88^{\pm.006}$

Table 3. Prompt V2 : Changes in uncertainty scores of the generated answers. The baseline represents the uncertainty when prompted with the question and giving the correct answer.

		Correct Answers			Incorrect Answers		
		H ↓	PPL ↓	TP ↑	H ↓	PPL ↓	TP ↑
GPT-4o	B	$0.08^{\pm.005}$	$1.05^{\pm.003}$	$0.97^{\pm.002}$	$0.47^{\pm.106}$	$1.42^{\pm.136}$	$0.80^{\pm.045}$
	FIP	$0.11^{\pm.005}$	$1.07^{\pm.004}$	$0.95^{\pm.002}$	$0.23^{\pm.030}$	$1.23^{\pm.109}$	$0.90^{\pm.017}$
	RIP	$0.13^{\pm.005}$	$1.07^{\pm.003}$	$0.95^{\pm.002}$	$0.45^{\pm.045}$	$1.34^{\pm.059}$	$0.83^{\pm.018}$
GPT-3.5	B	$0.17^{\pm.006}$	$1.09^{\pm.004}$	$0.93^{\pm.003}$	$0.44^{\pm.076}$	$1.35^{\pm.123}$	$0.83^{\pm.029}$
	FIP	$0.18^{\pm.006}$	$1.09^{\pm.004}$	$0.93^{\pm.003}$	$0.34^{\pm.028}$	$1.23^{\pm.028}$	$0.87^{\pm.011}$
	RIP	$0.19^{\pm.007}$	$1.10^{\pm.005}$	$0.93^{\pm.003}$	$0.48^{\pm.036}$	$1.34^{\pm.039}$	$0.81^{\pm.016}$
Mistral-7B	B	$0.22^{\pm.010}$	$1.12^{\pm.007}$	$0.92^{\pm.004}$	$0.71^{\pm.112}$	$1.63^{\pm.129}$	$0.69^{\pm.051}$
	FIP	$0.24^{\pm.012}$	$1.12^{\pm.009}$	$0.92^{\pm.005}$	$0.26^{\pm.019}$	$1.16^{\pm.018}$	$0.91^{\pm.008}$
	RIP	$0.24^{\pm.011}$	$1.13^{\pm.007}$	$0.92^{\pm.004}$	$0.47^{\pm.031}$	$1.30^{\pm.027}$	$0.82^{\pm.013}$
LLaMA-2-13B	B	$0.16^{\pm.006}$	$1.08^{\pm.004}$	$0.94^{\pm.003}$	$0.31^{\pm.017}$	$1.18^{\pm.014}$	$0.89^{\pm.007}$
	FIP	$0.20^{\pm.009}$	$1.10^{\pm.006}$	$0.93^{\pm.004}$	$0.23^{\pm.008}$	$1.12^{\pm.006}$	$0.92^{\pm.004}$
	RIP	$0.21^{\pm.007}$	$1.11^{\pm.006}$	$0.93^{\pm.003}$	$0.36^{\pm.013}$	$1.21^{\pm.012}$	$0.87^{\pm.006}$

argue that this happens because the incorrect answers likely reflect the incorrect information embedded in the FIP, pushing the model to be "overconfident" to distill fake information. Answers obtained with RIP generally exhibit high uncertainty scores since the given random information has nothing to do with the question, confusing the model even more. This effect is visible especially in incorrect answers, though also present with correct answers. At the same time,

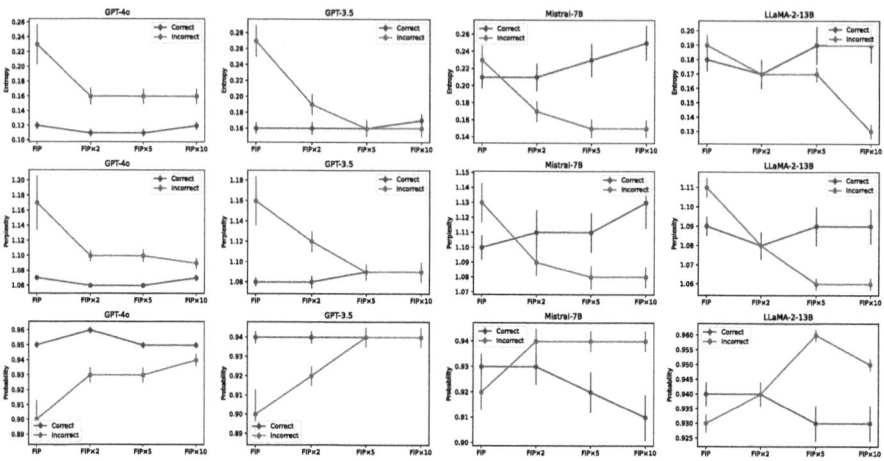

Fig. 2. Prompt V1 : Changes in uncertainty when repeating the false information ×k.

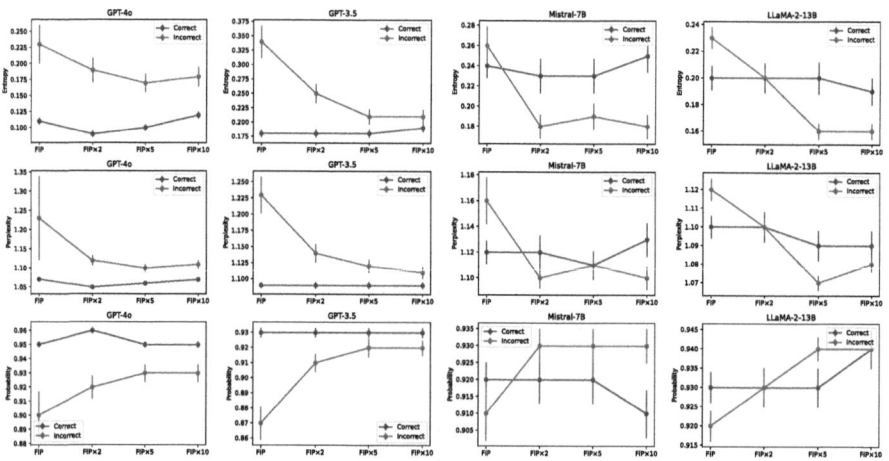

Fig. 3. Prompt V2 : Changes in uncertainty when repeating the false information ×k.

it does not lead to major drops in accuracy for the same reason, i.e., the prompt is not designed to target knowledge associated with the question.
Surprisingly, Mistral-7B is the only model that adheres to our expectations of not having any incorrect answers when prompted multiple times, contrary to the other ones (see Table 2 and the baseline in Table 4). Alas, this is not reported on the second prompt, which makes us believe Mistral to be more stable in its answers.

Influence of Repeated False Information on Model Confidence and Accuracy. To assess the impact of false information reoccurrence on the uncertainty levels of

Table 4. Correctness of the given answers after prompt manipulation (FIP/RIP) for both prompt types. We report average accuracies over 10 runs. Note that accuracies here refer to the subset of questions answered correctly – see Table 1.

	GPT-4o		GPT-3.5		Mistral-7B		LLaMA-2-13B	
	Prompt V1	Prompt V2	Prompt V1	Prompt V2	Prompt V1	Prompt V2	Prompt V1	Prompt V2
B	0.987	0.986	0.982	0.971	1.000	0.984	0.829	0.815
RIP	0.958	0.940	0.914	0.908	0.866	0.846	0.734	0.706
FIP	0.921	0.934	0.781	0.863	0.516	0.539	0.359	0.364
FIP×2	0.759	0.853	0.642	0.739	0.352	0.376	0.231	0.269
FIP×5	0.710	0.820	0.592	0.678	0.287	0.304	0.182	0.203
FIP×10	0.687	0.810	0.578	0.671	0.265	0.301	0.158	0.177
% FIP×10 vs. B	−30.4%	−17.8%	−41.1%	−30.9%	−73.5%	−69.4%	−80.9%	−78.3%

each model, we perform an ablation study where we repeat the FIP ×k with $k \in \{1, 2, 5, 10\}$. This means, the prompt includes the false information snippet k times, and is then followed by the question itself. Figures 2 and 3 illustrate this effect for Prompt V1 and V2, respectively. Notice the general trend for all models on the uncertainty of the incorrect answers when k increases: all models become consistently less uncertain about their incorrect answers (drops in entropy and perplexity, increase in token probability). Meanwhile, the uncertainty about the correct answers hits a plateau. When prompted multiple times with false knowledge, we suspect that the models become convinced of the new, contradicting information presented to them, hence their uncertainty about giving incorrect responses drops. Similarly, we want to see the effect of FIP on the overall accuracy of the models w.r.t. the baseline B. Table 4 shows this phenomenon for both prompt versions. As expected, the accuracy of correctly answering the given questions degrades abruptly when k increases. This is mostly emphasized for LLaMA-2-13B, reporting a degradation of −80.9% on FIP×10. Interestingly, the accuracy degradation for Prompt V2 is lower than that of V1. We argue that this happens due to specifically prompting the models to respond *truthfully* here.

Uncertainty Levels as Indicators for Adversarial Attacks. In our setting, the FIPs can be framed as adversarial prompts, as they aim to shift the previous, correct behaviour of the models. It is intuitive to assume rising uncertainty levels to be indicative for such adversarial attacks, i.e. sudden higher confusion in the model should indicate the model is currently being manipulated. While this can hold for single infusions of false information, uncertainty instead decreases the more false information is being presented, convincing the model. Hence, we argue uncertainty to not be a suitable tool for adversarial attack detection, calling for different methods and rendering this potential approach inappropriate.

4 Conclusion

In this study, we explored the knowledge drift of GPT-4o, GPT-3.5, Mistral-7B, and LLaMA-2-13B through the impact of false information on their performance and uncertainty within a Question Answering setting using the TriviaQA dataset. Our findings reveal that presenting false information to the models successfully introduces knowledge drift in their responses, as well as generally increases the responses' uncertainty. Notably however, repeated exposure to the same false prompts can convince the model of the given false information, leading it to be more certain of the incorrect answers. Additionally, we observed that random and unrelated information results in the highest uncertainty, highlighting the model's greater confusion with irrelevant and noisy data.

Our findings provide valuable insights to the drift that is possible in LLMs' internal knowledge structures, and underscore the complexities in the knowledge processing of LLMs as well as its reflections in their uncertainty levels. With this, we aim to contribute to the broader effort to improve the robustness and reliability of language models, particularly in the face of adversarial inputs. Understanding and enhancing their resilience remains imperative as language models become increasingly integrated into critical applications.

In the future, we will explore these dynamics across different datasets and develop advanced techniques to enhance LLMs' robustness and trustworthiness further. An especially interesting extension of this study would be to infuse the false information into the model through continued training on the false data instead of presenting the false information during inference in the form of prompting, since this approach does effectively not change anything about the model's internal knowledge. Lastly, we aim to incorporate an "adversarial protection" mechanism into state-of-the-art models to ensure their effective and safe deployment in real-world scenarios.

References

1. Achiam, J., et al.: Gpt-4 technical report. arXiv preprint arXiv:2303.08774 (2023)
2. Brown, T.B., et al.: Language models are few-shot learners. Adv. Neural. Inf. Process. Syst. **33**, 1877–1901 (2020)
3. Chen, C., et al.: Inside: Llms' internal states retain the power of hallucination detection. arXiv preprint arXiv:2402.03744 (2024)
4. Chowdhery, A., et al.: Palm: scaling language modeling with pathways. J. Mach. Learn. Res. **24**(240), 1–113 (2023)
5. Desai, S., Durrett, G.: Calibration of pre-trained transformers. arXiv preprint arXiv:2003.07892 (2020)
6. Devlin, J., Chang, M., Lee, K., Toutanova, K.: BERT: pre-training of deep bidirectional transformers for language understanding. In: Burstein, J., Doran, C., Solorio, T. (eds.) Proceedings of the 2019 Conference of the North American Chapter of the Association for Computational Linguistics: Human Language Technologies, NAACL-HLT 2019, Minneapolis, MN, USA, June 2-7, 2019, Volume 1 (Long and Short Papers). pp. 4171–4186. Association for Computational Linguistics (2019). https://doi.org/10.18653/V1/N19-1423
7. Glushkova, T., Zerva, C., Rei, R., Martins, A.F.: Uncertainty-aware machine translation evaluation. arXiv preprint arXiv:2109.06352 (2021)
8. Huang, L., et al.: A survey on hallucination in large language models: Principles, taxonomy, challenges, and open questions. arXiv preprint arXiv:2311.05232 (2023)

9. Jiang, Z., Araki, J., Ding, H., Neubig, G.: How can we know When language models know? on the calibration of language models for question answering. Trans. Assoc. Comput. Linguisti. **9**, 962–977 (2021). https://doi.org/10.1162/TACL_A_00407
10. Joshi, M., Choi, E., Weld, D.S., Zettlemoyer, L.: Triviaqa: A large scale distantly supervised challenge dataset for reading comprehension. arXiv preprint arXiv:1705.03551 (2017)
11. Kadavath, S., et al.: Language models (mostly) know what they know. arXiv preprint arXiv:2207.05221 (2022)
12. Kuhn, L., Gal, Y., Farquhar, S.: Semantic uncertainty: Linguistic invariances for uncertainty estimation in natural language generation. In: The Eleventh International Conference on Learning Representations, ICLR 2023, Kigali, Rwanda, May 1-5, 2023. OpenReview.net (2023)
13. Lin, S., Hilton, J., Evans, O.: Teaching models to express their uncertainty in words. Trans. Mach. Learn. Res. **2022** (2022)
14. Mielke, S.J., Szlam, A., Dinan, E., Boureau, Y.L.: Reducing conversational agents' overconfidence through linguistic calibration. Transact. Assoc. Comput. Linguist. **10**, 857–872 (2022)
15. Ott, M., Auli, M., Grangier, D., Ranzato, M.: Analyzing uncertainty in neural machine translation. In: International Conference on Machine Learning, pp. 3956–3965. PMLR (2018)
16. Park, P.S., Goldstein, S., O'Gara, A., Chen, M., Hendrycks, D.: Ai deception: A survey of examples, risks, and potential solutions. Patterns **5**(5) (2024)
17. Petroni, F., et al.: Language models as knowledge bases? arXiv preprint arXiv:1909.01066 (2019)
18. Radford, A., Narasimhan, K., Salimans, T., Sutskever, I., et al.: Improving language understanding by generative pre-training (2018)
19. Raffel, C., et al.: Exploring the limits of transfer learning with a unified text-to-text transformer. J. Mach. Learn. Res. **21**(140), 1–67 (2020)
20. Ranaldi, L., Pucci, G.: When large language models contradict humans? large language models' sycophantic behaviour. arXiv preprint arXiv:2311.09410 (2023)
21. Roberts, A., Raffel, C., Shazeer, N.: How much knowledge can you pack into the parameters of a language model? In: Webber, B., Cohn, T., He, Y., Liu, Y. (eds.) Proceedings of the 2020 Conference on Empirical Methods in Natural Language Processing, EMNLP 2020, Online, November 16-20, 2020. pp. 5418–5426. Association for Computational Linguistics (2020). https://doi.org/10.18653/V1/2020.EMNLP-MAIN.437
22. Sai, A.B., Mohankumar, A.K., Khapra, M.M.: A survey of evaluation metrics used for nlg systems. ACM Comput. Surv. (CSUR) **55**(2), 1–39 (2022)
23. Touvron, H., et al.: Llama: Open and efficient foundation language models. arXiv preprint arXiv:2302.13971 (2023)
24. Wang, Y., Beck, D., Baldwin, T., Verspoor, K.: Uncertainty estimation and reduction of pre-trained models for text regression. Transact. Assoc. Comput. Linguist. **10**, 680–696 (2022)
25. Youssef, P., Koras, O.A., Li, M., Schlötterer, J., Seifert, C.: Give me the facts! A survey on factual knowledge probing in pre-trained language models. In: Bouamor, H., Pino, J., Bali, K. (eds.) Findings of the Association for Computational Linguistics: EMNLP 2023, Singapore, December 6-10, 2023, pp. 15588–15605. Association for Computational Linguistics (2023). https://doi.org/10.18653/V1/2023.FINDINGS-EMNLP.1043

Exploring Concept Drift Visualization and Explanation in Image Streams

Giacomo Ziffer[(✉)] and Emanuele Della Valle

DEIB, Politecnico di Milano, Milano, Italy
{giacomo.ziffer,emanuele.dellavalle}@polimi.it

Abstract. In the rapidly evolving field of Machine Learning applied to data streams, where information arrives continuously and models must adapt in real-time, drift detection has emerged as a critical component for maintaining model accuracy and reliability. While much attention has been given to structured data, the challenge of handling real-time image streams remains comparatively underexplored, despite its vital role in numerous applications. This study addresses this gap by exploring innovative approaches to visualize and explain drift in image streams. We propose a novel method that leverages zero-shot classification capabilities of a pre-trained ResNet-50 architecture to visualize potential changes within the stream. Additionally, we investigate the application of ResNet-50 in combination with Uniform Manifold Approximation and Projection. This hybrid approach aims to identify and visualize drift in high-dimensional image data by projecting it onto a lower-dimensional space, facilitating easier interpretation and analysis of evolving patterns over time. Preliminary results on the CLEAR10 dataset, the first benchmark with a natural temporal evolution of visual concepts, showcase potential applications for drift monitoring in image streams. These findings present a promising avenue for further research and development, aiming at solidifying the integration of these methodologies for improved drift detection and overall models performance in dynamic data contexts.

Keywords: Image classification · Data streams · Concept drift

1 Introduction

Continuous learning is increasingly recognized in Machine Learning and Artificial Intelligence, especially with the integration of sensor networks and the Internet of Things. Real-time applications demand immediate online learning for each instance, posing challenges in processing entire datasets due to resource constraints. Streaming Machine Learning (SML) has emerged as a pivotal paradigm [1]. SML operates under unique constraints: (i) single-pass learning, observing each item once; (ii) minimal processing time per item; (iii) constrained memory usage, ideally sub-linear with stream length; (iv) real-time provision of answers; and (v) adaptation to dynamic data sources' evolution over time.

While the SML community has extensively explored classifiers for simpler data streams [13], there is a notable gap in addressing challenges posed by complex and high-dimensional data streams [5]. SML research has primarily focused on tabular datasets due to computational constraints and the need for efficient single-pass processing in highly non-stationary scenarios. With fewer features, these datasets serve as foundational testing grounds for SML algorithms. However the application of these methods to high multidimensional data such as image streams still remains under-exploited, with only few works investigating the problem [6,9,19]. This represents a significant gap, given the importance of images in many modern applications such as surveillance, autonomous driving, and medical diagnostics. However, image streams with intricate spatio-temporal dependencies require more sophisticated methodologies for visualizing and detecting evolving patterns. Indeed, a further significant challenge in learning from data streams is that they are typically temporally arranged [27], with changes in the distribution resulting from the passage of time. In the context of time-ordered image streams, the temporal variation in the data distribution is formally defined as temporal distribution shift [26].

Deep neural networks (DNNs) are the prevailing approach in image classification, particularly for tasks involving semantic extraction from sensor data such as cameras [20]. However, conventional DNNs face challenges in adapting to incremental updates or rapidly learning from individual instances due to the stability-plasticity dilemma [15]. Recent research has largely focused on incremental DNNs capable of addressing catastrophic forgetting [10,14]; yet, most of these algorithms operate within an incremental batch learning framework. This presents an opportunity to further explore the problem from a SML perspective, utilizing DNNs in inference-only mode to effectively extract low-dimensional features from image streams.

In this preliminary study, we propose two methodologies. The first approach utilizes zero-shot classification with pre-trained DNNs to study evolving patterns in the image stream. The second one leverages pre-trained DNNs to extract low-dimensional features from images, which are then visualized using dimensionality reduction techniques such as Uniform Manifold Approximation and Projection (UMAP) [16]. We conduct preliminary experiments on the CLEAR10 dataset [11], the first benchmark with a natural temporal evolution of visual concepts, showcasing potential applications for drift monitoring in image streams. All data streams and algorithms used in this paper are available online[1] as a publicly available benchmark to help other researchers to reproduce the results shown in this study or the development of new algorithms.

The remainder of this paper is organized as follows. Section 2 presents relevant works on concept drift detection, explanation, and visualization. Section 3 provides a detailed overview of the proposed methodologies, while Sect. 4 introduces the CLEAR10 dataset and discusses the initial experimental results. Finally, Sect. 5 summarizes our findings and outlines future research directions for investigating concept drift in the context of image streams.

---

[1] https://github.com/gziffer/image-stream-drift-analysis.

2 Related Works

The phenomenon of concept drift, characterized by changes in the statistical properties of data over time, has been extensively studied within the area of Machine Learning, particularly in the context of streaming data. The research community has primarily focused on strategies for detecting and adapting to concept drift, emphasizing the critical importance of timely detection to maintain model accuracy [3,4,13]. As machine learning models are deployed in dynamic environments, their performance can degrade if they fail to adapt to these changes. This necessitates a robust understanding of concept drift's underlying causes and mechanisms.

In recent years, explainability has emerged as a significant area of research, particularly regarding the need for models to provide interpretable outputs [18]. However, the development of more complex explanation methods specifically tailored for concept drift remains limited. Various approaches have been proposed to detect and quantify drift [13,22,24], localize it in space [7,12,13,22], and visualize its effects using feature-wise representations of drift [17,23,24]. Nevertheless, these techniques often encounter challenges when dealing with high-dimensional data or non-semantic features, highlighting the need for further research.

A notable contribution to the understanding of concept drift explored model-based explanations of concept drift [8]. The authors proposed a methodology that simplifies the explanation of concept drift by framing it in terms of models trained to extract relevant information regarding the drift. This approach facilitates the use of a diverse array of explanation schemes, allowing researchers to select the most appropriate method for addressing specific drift-related challenges. By characterizing concept drift through the changes in spatial features using various explanation techniques, the authors aimed to provide human-understandable descriptions of how and where drift manifests, ultimately enhancing the acceptance of lifelong learning models.

Only few works concentrated on the application of concept drift detection and visualization techniques to image streams, as most studies have focused on structured data. Yang et al. [25] discussed visual analytics approaches for diagnosing concept drift, suggesting that traditional drift detection methods can be effectively adapted for image classification tasks. However, the challenges posed by high-dimensional unstructured data streams are still largely unexplored. This area requires focused research to address the evolution of image features over time and the resulting impact on the accuracy of streaming classifiers.

3 Proposed Methods

In this section, we present the methodologies proposed for conducting drift analysis in image streams. The approach involves leveraging pre-trained neural networks for feature extraction, followed by the application of UMAP as the chosen technique for dimensionality reduction. This combination aims to effectively capture and analyze changes in the underlying data distribution over time.

Zero-Shot Classification for Drift Visualization. The first approach leverages zero-shot classification using pre-trained deep neural networks (DNNs) to monitor and visualize drift in image streams. In this method, a neural network trained on a broad and diverse dataset, such as ImageNet, is employed in an inference-only mode without any additional fine-tuning for the specific data stream being analyzed. This design choice capitalizes on the generalization capability of the pre-trained model, allowing it to recognize and classify a wide range of image categories even when faced with new and previously unseen data.

The pre-trained DNN is directly applied to classify incoming images, and the resulting predicted classes are continuously analyzed to observe how their distribution evolves over time. By tracking changes in the distribution of these predictions, we can identify potential drift within the data stream. Since the model operates without fine-tuning, its predictions are based solely on its initial training, making any shifts in class distribution particularly valuable for detecting drift. By cross-referencing the predicted classes with those in the model's original training dataset, we can assess whether the model is encountering familiar or entirely new categories. This capability is crucial for understanding if the model can recognize classes independently of any further training.

This approach is especially effective for identifying drifts that involve the emergence of new classes or the reappearance of previously unseen ones within the data stream. The pre-trained DNN's ability to generalize across a wide spectrum of categories provides a direct and efficient method for tracking changes in data distribution, making it highly useful in scenarios where quickly detecting shifts in data patterns is critical. By utilizing the model in an inference-only mode, the approach remains robust to new data while avoiding the complexities and computational overhead associated with fine-tuning. This ensures that the model can adapt to evolving data streams without compromising its performance, offering a scalable and practical solution for real-time drift detection.

Feature Extraction and UMAP for Drift Analysis. The second approach involves feature extraction from a pre-trained DNN, but instead of using the final classification layer, we focus on the intermediate embeddings. These embeddings represent low-dimensional features that capture the essential characteristics of the images. After extracting these features, we apply Uniform Manifold Approximation and Projection (UMAP) [16] for dimensionality reduction. UMAP is chosen for its ability to maintain the temporal orientation of the data, which is crucial for understanding the evolution of patterns over time.

This method leverages the robust feature extraction capabilities of deep learning and the efficiency and interpretability of UMAP. By visualizing the reduced-dimensional embeddings, we can detect and visualize distributional shifts in the feature space. Additionally, this methodology can be extended by integrating distance metrics in the UMAP embedding space, such as the Wasserstein distance, to quantify the variation in the position of data points. This approach can be used both for real-time drift detection and for exploratory analysis of potential drifts in image streams.

Fig. 1. Display of the temporal evolution of visual concepts Baseball, Bus, Camera, Cosplay, Dress, Hockey, Laptop, Racing, Soccer, Sweater in CLEAR10 dataset [11].

4 Experimental Evaluation

4.1 CLEAR10 Dataset

CLEAR10, a benchmark designed for continuous learning of real-world images, introduces a novel approach by focusing on the natural temporal evolution of visual concepts in Internet images. Leveraging the YFCC100M [21] dataset and timestamps of images spanning 2004 to 2014, CLEAR10 organizes the data into a temporal stream divided into 11 'buckets'. Each bucket, representing a specific time period, features a subset of 11 temporally dynamic classes. These classes include 10 illustrative categories (e.g., baseball, bus, camera, cosplay, dress, hockey, laptop, racing, soccer, sweater) and an additional background class, with 300 labeled images per class, as shown in Fig. 1.

Analyzing the evolution of visual concepts within the CLEAR10 benchmark spanning 2004 to 2014 offers a compelling perspective for applying adaptive SML classifiers. The dynamic nature of visual content over this timeframe encapsulates technological advancements and shifting trends, providing an intricate stream of data. From the transition in camera technologies, exemplified by the progression from Canon EOS 30D to Canon EOS 6D, to the transformation of the term "camera" itself (from traditional stand-alone devices to integration in smart-

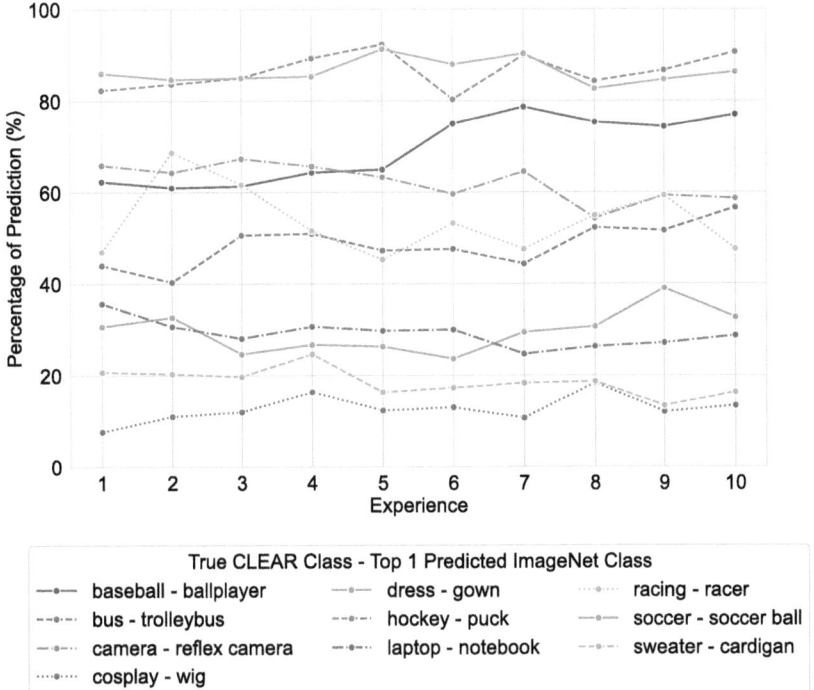

Fig. 2. Evolution of top-1 prediction percentages for CLEAR classes. The graph shows the highest prediction confidence for each class across 10 buckets. The legend shows each CLEAR class name followed by its corresponding predicted ImageNet label.

phones like the iPhone 5 [11]) these changes pose unique challenges for real-time learning systems. Consequently, the CLEAR10 benchmark becomes a valuable testbed, as it necessitates continuous adaptation to evolving visual landscapes, reflecting the real-world challenges faced by systems operating in dynamic and temporally evolving environments.

4.2 Results

The analysis firstly investigates zero-shot transfer learning capabilities using visual features extracted from a ResNet-50 model pre-trained on ImageNet [2]. This approach enables us to evaluate how well ImageNet-derived features generalize to CLEAR classes without any additional training. Figure 2 and Fig. 3 illustrate the temporal evolution of these zero-shot predictions across the ten CLEAR chronologically ordered data buckets. Figure 2 tracks the highest confidence predictions (top-1) for each CLEAR class, while Fig. 3 presents the second-highest confidence predictions (top-2). Together, these visualizations provide insights into both the strength and stability of the cross-dataset class relationships over time.

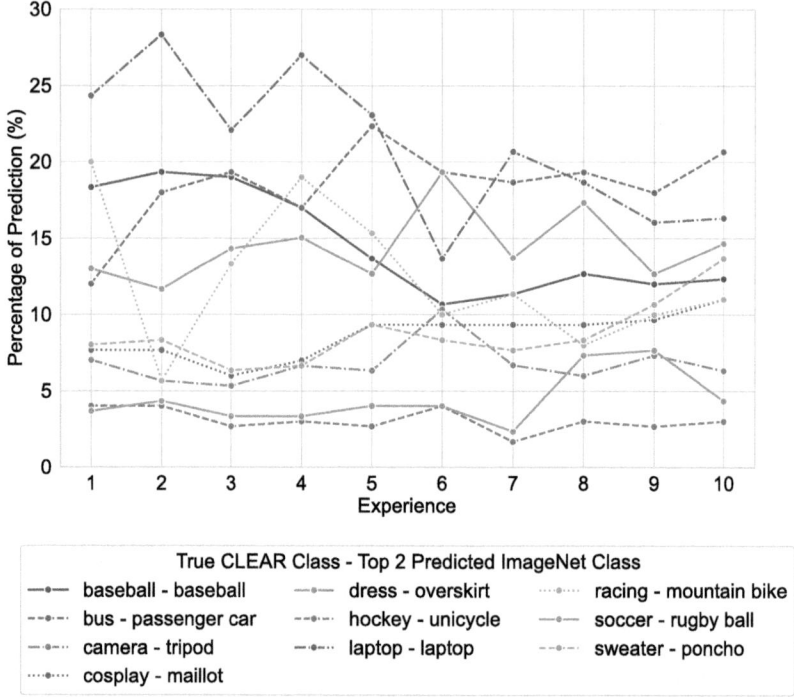

Fig. 3. Evolution of top-2 prediction percentages for CLEAR classes. The graph shows the second highest prediction confidence for each class across 10 buckets. The legend shows each CLEAR class name followed by its corresponding predicted ImageNet label.

In Fig. 2, we observe distinct patterns in prediction confidence across different classes. Two CLEAR classes exhibit remarkably high prediction confidence throughout all experiences: hockey and soccer consistently maintain percentages above 80%, with hockey reaching peaks around 90%. These classes demonstrate particularly strong and stable associations with their corresponding ImageNet categories, i.e., puck and soccer ball, respectively. The medium confidence range (40–80%) includes classes like camera and soccer, which show more variable patterns but maintain relatively consistent prediction levels. Notably, baseball displays a gradual increase in prediction confidence from around 60% to nearly 80% across experiences, suggesting a strengthening of the class association over time. The lower confidence range (below 40%) includes classes like cosplay, which shows more volatile prediction patterns, indicating less stable mappings to ImageNet categories. This stratification in prediction confidence suggests also varying degrees of semantic alignment between CLEAR and ImageNet class definitions.

Figure 3, depicting the second-most confident predictions, reveals substantially lower percentage values, all below 30%. This marked difference between top-1 and top-2 prediction confidences suggests the presence of strong, singular associations rather than distributing probability mass across multiple classes. The temporal evolution of predictions shows interesting patterns across years. While bus and dress maintain stable confidence predictions, laptop shows more variable behavior with noticeable fluctuations. This suggests ongoing refinement of certain class relationships over the years, potentially reflecting changes in the underlying data distribution or the model's ability to identify subtle class-specific features across different experiences. The sweater class demonstrates a particularly interesting trend, showing a gradual improvement in prediction confidence over experiences, indicating progressive refinement of its class representation.

Figure 3 show more erratic patterns compared to Fig. 2. For instance, laptop and baseball classes display crossover patterns in their secondary predictions, suggesting less stable secondary associations between these CLEAR classes and their corresponding ImageNet categories. This behavior is particularly evident in experiences 4 through 7, where several classes show increased variability in their second-choice predictions. These findings suggest that while the model develops strong primary associations between CLEAR and ImageNet classes, particularly for visually distinct categories like camera and dress, the relationship is not uniformly strong across all classes. The varying levels of prediction confidence and stability between classes like notebook and laptop compared to camera and dress could indicate inherent differences in the visual distinctiveness of these categories or varying degrees of semantic overlap between CLEAR and ImageNet class definitions.

Further evidence supporting these findings is presented in Fig. 4, which provides an aggregate visualization of the prediction patterns previously detailed in Fig. 2 and Fig. 3. This comprehensive visualization maps all the relationship between CLEAR and ImageNet classes across temporal buckets. The x-axis represents CLEAR10 classes (including the background class), subdivided by temporal buckets, while the y-axis displays ImageNet classes, starting from class 400 as no significant predictions were observed below this threshold. The visualization employs a bubble chart format where circle diameter and color intensity encode the prediction frequency: larger, darker circles indicate higher frequencies of a particular ImageNet class being predicted for the corresponding CLEAR10 images. This aggregated view reinforces the above observations about prediction stability and class relationships while providing additional insights into the broader distribution of predictions across the ImageNet class space.

Several insights can be drawn from these results. Three CLEAR10 classes, baseball, hockey, and soccer, have a predominant ImageNet class, with over 250 predictions per bucket out of 300 images per class per bucket. This suggests that for these three CLEAR10 classes, a single ImageNet class maps very well, and the lack of significant variation over the years indicates that these classes have mostly stayed the same. Notably, we are using a model that is not continuously learning, so we can evaluate how patterns change relative to a pre-learned pattern.

Moreover, a notable observation pertains to the Background class in the first column. As expected, no dominant class emerges, suggesting ongoing changes in these images over the years. Another challenging category is cosplay, which

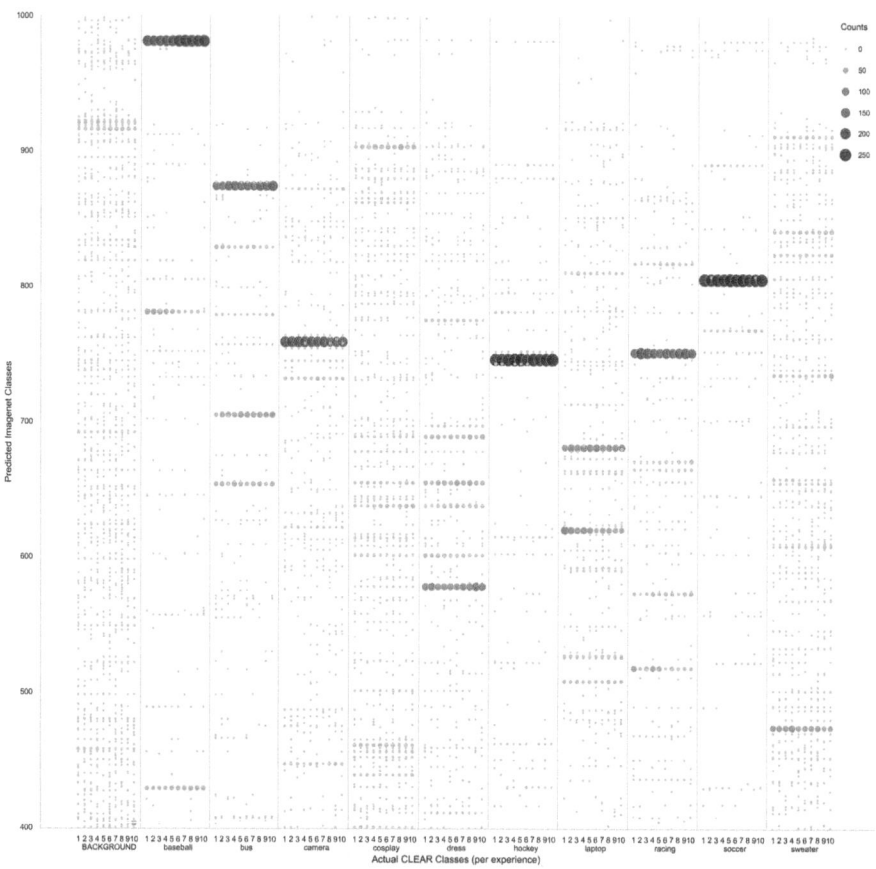

Fig. 4. Visualization of the CLEAR10 classes using a Resnet50 pretrained on the ImageNet dataset.

is absent from ImageNet and frequently categorized as various types of clothing. Particularly, this class is often predicted as the same ImageNet class as CLEAR10 dress class, indicating a notable pattern possibly indicative of gradual and recurrent drift. This is supported by the fluctuation in the number of predictions over time, implying a resemblance among non-consecutive images.

Finally, Fig. 5 depicts the embedding visualization of the cosplay class across the ten buckets. In this scenario, UMAP was initially fitted using data from the first bucket and subsequently applied to data from subsequent years, mimicking a real-world scenario where the feature extraction pipeline is updated only when drift is detected. This visualization provides insightful patterns in the learned feature space distribution. Each class exhibits distinct and meaningful spatial arrangements that persist across temporal buckets, revealing several interesting characteristics of the feature representations.

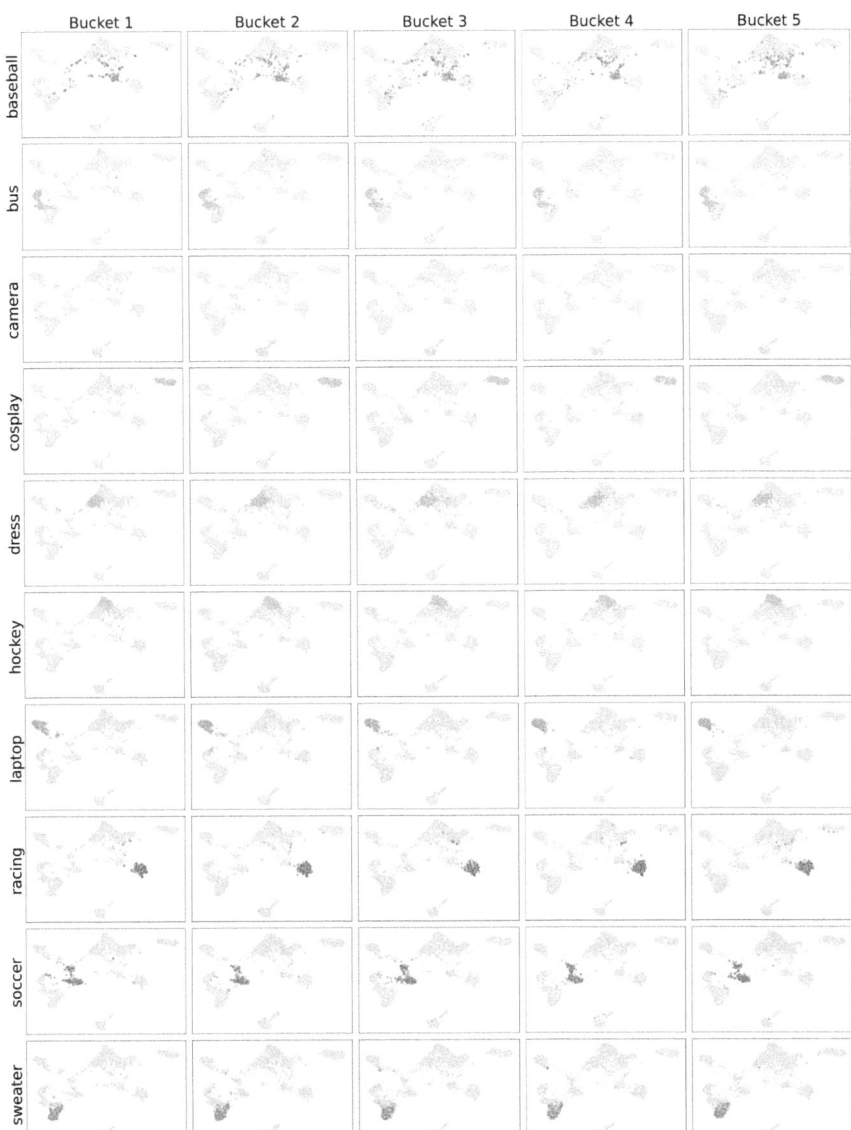

Fig. 5. UMAP visualization of the feature space for each CLEAR class (rows) across the first five temporal buckets (columns). Each subplot displays the distribution of image features in a two-dimensional space, with the target class highlighted in color against the grey background of all other classes. (Color figure online)

It is noteworthy how visually complex classes achieve clear separation in the feature space. For instance, cosplay images, despite their inherent visual complexity and variability as shown in previous analysis, form a well-defined cluster that occupies a unique region in the feature space, distinctly separated from other classes. This spatial isolation suggests that the latent space has successfully captured the unique visual characteristics that define this category. Similar effective separation is observed for racing class samples, which consistently cluster in a specific region of the feature space across all temporal buckets. This stable, isolated clustering indicates robust feature extraction for racing-related visual elements, despite the dynamic nature of racing scenes. Sports-related classes like "soccer" and "hockey" show distinctive clustering patterns while maintaining some proximity in the feature space, suggesting that the model captures both their unique characteristics and their shared sports-related features. The "baseball" class exhibits a more dispersed yet consistent distribution pattern, reflecting the variety of baseball-related scenes while maintaining class coherence.

The preservation of these spatial relationships across temporal buckets indicates the quality of the feature representations learned by the model. This temporal consistency suggests that the underlying visual features remain stable and discriminative over time, thereby facilitating reliable class identification.

5 Conclusion

This preliminary study underscores the importance of drift analysis in image streams, presenting two approaches for monitoring and detecting distributional shifts while enhancing explainability. By leveraging pre-trained ResNet-50 deep neural networks, our methodologies extract more meaningful features, rendering them more conducive to thorough analysis and interpretation. Conversely, the feature extraction and UMAP visualization approach facilitate the visualization of distribution shifts in the feature space, establishing a robust framework for drift analysis in streaming image data. Our exploration of the CLEAR10 dataset yields invaluable insights into the evolving nature of visual concepts over time, thereby elucidating the challenges inherent in non-stationary image streams.

Future research should focus on comprehensive validation of these methods through rigorous testing. A promising direction is adapting the UMAP framework for drift detection in image streams. Since the feature space and inter-point distances are already computed, this could enable straightforward drift detection by monitoring distance variations over time. Further investigation could explore combining this approach with statistical change point detection methods to improve drift sensitivity while maintaining low false alarm rates.

References

1. Bifet, A., Gavaldà, R., Holmes, G., Pfahringer, B.: Machine Learning for Data Streams: With Practical Examples in MOA. MIT press (2018)
2. Deng, J., Dong, W., Socher, R., Li, L., Li, K., Fei-Fei, L.: ImageNet: a large-scale hierarchical image database. In: CVPR, pp. 248–255. IEEE Computer Society (2009)
3. Ditzler, G., Roveri, M., Alippi, C., Polikar, R.: Learning in nonstationary environments: a survey. IEEE Comput. Intell. Mag. **10**(4), 12–25 (2015)
4. Gama, J., Zliobaite, I., Bifet, A., Pechenizkiy, M., Bouchachia, A.: A survey on concept drift adaptation. ACM Comput. Surv. **46**(4), 44:1–44:37 (2014)
5. Gomes, H.M., Read, J., Bifet, A., Barddal, J.P., Gama, J.: Machine learning for streaming data: state of the art, challenges, and opportunities. KDD **21**, 6–22 (2019)
6. Hayes, T.L., Kanan, C.: Lifelong machine learning with deep streaming linear discriminant analysis. In: CVPR Workshops, pp. 887–896. Computer Vision Foundation / IEEE (2020)
7. Hinder, F., Vaquet, V., Brinkrolf, J., Artelt, A., Hammer, B.: Localization of concept drift: identifying the drifting datapoints. In: IJCNN, pp. 1–9. IEEE (2022)
8. Hinder, F., Vaquet, V., Brinkrolf, J., Hammer, B.: Model-based explanations of concept drift. Neurocomputing **555**, 126640 (2023)
9. Korycki, Ł., Krawczyk, B.: Streaming decision trees for lifelong learning. In: Oliver, N., Pérez-Cruz, F., Kramer, S., Read, J., Lozano, J.A. (eds.) ECML PKDD 2021. LNCS (LNAI), vol. 12975, pp. 502–518. Springer, Cham (2021). https://doi.org/10.1007/978-3-030-86486-6_31
10. Lesort, T., Lomonaco, V., Stoian, A., Maltoni, D., Filliat, D., Rodríguez, N.D.: Continual learning for robotics: definition, framework, learning strategies, opportunities and challenges. Inf. Fusion **58**, 52–68 (2020)
11. Lin, Z., Shi, J., Pathak, D., Ramanan, D.: The clear benchmark: continual learning on real-world imagery. In: Thirty-fifth Conference on Neural Information Processing Systems Datasets and Benchmarks Track (2021)
12. Liu, A., Song, Y., Zhang, G., Lu, J.: Regional concept drift detection and density synchronized drift adaptation. In: IJCAI, pp. 2280–2286 (2017). https://www.ijcai.org/
13. Lu, J., Liu, A., Dong, F., Gu, F., Gama, J., Zhang, G.: Learning under concept drift: a review. IEEE Trans. Knowl. Data Eng. **31**(12), 2346–2363 (2019)
14. Mai, Z., Li, R., Jeong, J., Quispe, D., Kim, H., Sanner, S.: Online continual learning in image classification: an empirical survey. Neurocomputing **469**, 28–51 (2022)
15. McCloskey, M., Cohen, N.J.: Catastrophic interference in connectionist networks: the sequential learning problem. In: Psychology of Learning and Motivation, vol. 24, pp. 109–165. Elsevier (1989)
16. McInnes, L., Healy, J.: UMAP: uniform manifold approximation and projection for dimension reduction. CoRR abs/1802.03426 (2018)
17. Pratt, K.B., Tschapek, G.: Visualizing concept drift. In: KDD, pp. 735–740. ACM (2003)
18. Rohlfing, K.J., et al.: Explanation as a social practice: toward a conceptual framework for the social design of AI systems. IEEE Trans. Cogn. Dev. Syst. **13**(3), 717–728 (2021)
19. Rožanec, J.M., Trajkova, E., Dam, P., Fortuna, B., Mladenić, D.: Streaming machine learning and online active learning for automated visual inspection. IFAC-PapersOnLine **55**(2), 277–282 (2022)

20. Sharma, N., Jain, V., Mishra, A.: An analysis of convolutional neural networks for image classification. Procedia Comput. Sci. **132**, 377–384 (2018). International Conference on Computational Intelligence and Data Science
21. Thomee, B., et al.: YFCC100M: the new data in multimedia research. Commun. ACM **59**(2), 64–73 (2016)
22. Wang, P., Yu, H., Jin, N., Davies, D., Woo, W.L.: QuadCDD: a quadruple-based approach for understanding concept drift in data streams. Expert Syst. Appl. **238**(Part E), 122114 (2024)
23. Wang, X., et al.: ConceptExplorer: Visual analysis of concept drifts in multi-source time-series data. In: IEEE VAST@IEEE VIS, pp. 1–11. IEEE (2020)
24. Webb, G.I., Lee, L.K., Goethals, B., Petitjean, F.: Analyzing concept drift and shift from sample data. Data Min. Knowl. Disc. **32**(5), 1179–1199 (2018). https://doi.org/10.1007/s10618-018-0554-1
25. Yang, W., et al.: Diagnosing concept drift with visual analytics. In: IEEE VAST@IEEE VIS, pp. 12–23. IEEE (2020)
26. Yao, H., Choi, C., Cao, B., Lee, Y., Koh, P.W., Finn, C.: Wild-time: A benchmark of in-the-wild distribution shift over time. In: NeurIPS (2022)
27. Ziffer, G., Bernardo, A., Della Valle, E., Cerqueira, V., Bifet, A.: Towards time-evolving analytics: online learning for time-dependent evolving data streams. Data Sci. **6**(1–2), 1–16 (2023)

Innovative Approaches to Concept Drift Detection and Landscape Shifts

A Synthetic Benchmark to Explore Limitations of Localized Drift Detections

Flavio Giobergia[1](✉), Eliana Pastor[1], Luca de Alfaro[2], and Elena Baralis[1]

[1] Politecnico di Torino, Turin, Italy
{flavio.giobergia,eliana.pastor,elena.baralis}@polito.it
[2] University of California, Santa Cruz, USA
luca@ucsc.edu

Abstract. Concept drift is a common phenomenon in data streams where the statistical properties of the target variable change over time. Traditionally, drift is assumed to occur globally, affecting the entire dataset uniformly. However, this assumption does not always hold true in real-world scenarios where only specific subpopulations within the data may experience drift. This paper explores the concept of localized drift and evaluates the performance of several drift detection techniques in identifying such localized changes. We introduce a synthetic dataset based on the Agrawal generator, where drift is induced in a randomly chosen subgroup. Our experiments demonstrate that commonly adopted drift detection methods may fail to detect drift when it is confined to a small subpopulation. We propose and test various drift detection approaches to quantify their effectiveness in this localized drift scenario. We make the source code for the generation of the synthetic benchmark available at https://github.com/fgiobergia/subgroup-agrawal-drift.

Keywords: Drift detection · Synthetic data · Localized drift

1 Introduction

In the realm of data stream mining, the detection of concept drift is of fundamental importance to maintain the accuracy and reliability of predictive models. Concept drift refers to the change in the statistical properties of the target variable that the model is trying to predict. Traditionally, drift detection techniques make the (often implicit) assumption that the drift occurs globally, i.e., the change is uniformly distributed across the entire dataset. This assumption, however, may not always hold in real-world situations where drift can occur in a localized manner, affecting only certain subpopulations within the data (e.g., only young women employed in the IT sector).

Localized drift poses a significant challenge for traditional drift detection methods. These methods are designed to identify global changes and may overlook drifts that are confined to a small subset of the data. As a result, models may fail to adapt to these local changes, leading to degraded performance and inaccurate predictions. For instance, a subgroup covering 2% of the population may start behaving in a significantly different way than previously known. It is desirable that this change in behavior be detected by drift detectors. However, the drift goes unnoticed when we observe the overall performance of the model (i.e., the performance on the entire population). Figure 1 shows how the accuracy varies under subgroup drift for the entire population and for the specific subgroup. While the overall performance degrades by approximately 2% and may go unnoticed, the accuracy within the subgroup drops to 0.

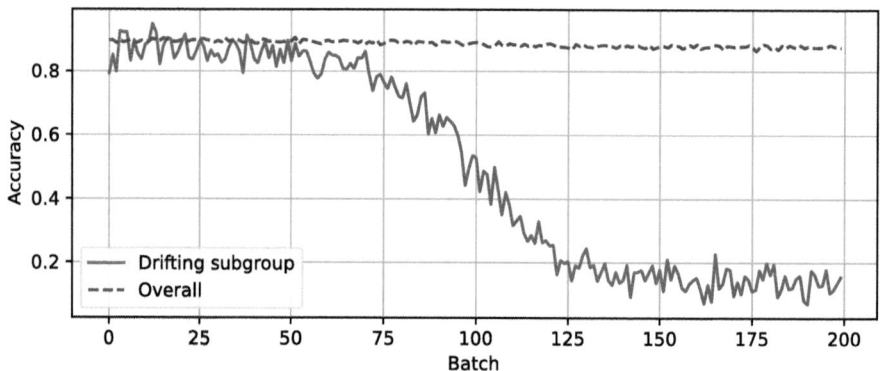

Fig. 1. Accuracy computed on the overall dataset and on the drifting subgroup (2% of the dataset), throughout a drifting event.

To investigate the limitations of existing drift detection methods in the context of localized drift, we introduce a synthetic dataset inspired by the Agrawal generator [1]. In this dataset, drift is intentionally induced in a randomly chosen subgroup of a specific size, while the rest of the data remains stable. This setup allows us to simulate a scenario where only a specific subpopulation is subject to drift, thereby providing a controlled environment to evaluate the effectiveness of various drift detection techniques.

The primary contributions of this paper are as follows:

- We highlight the importance of recognizing localized drift in data streams and its implications for drift detection methodologies.
- We introduce a synthetic dataset based on the Agrawal generator with induced localized drift, providing a benchmark for evaluating drift detection methods.
- We conduct a comprehensive evaluation of several drift detection techniques, quantifying their performance in detecting localized drift.

The rest of the paper is organized as follows. Section 2 describes the synthetic dataset and experimental setup. Section 3 presents the results of our experiments and discusses the findings. Finally, Sect. 4 concludes the paper and suggests directions for future research.

2 Proposed Dataset

We introduce a novel dataset based on the synthetic one proposed in [1]. In particular, we propose (i) identifying a randomly selected subgroup of the population, defined as a slice of the dataset's attributes and of a user-specified size, and (ii) only injecting this target subgroup with noise to simulate a situation where the drift occurs locally, instead of globally. The code is available at https://github.com/fgiobergia/subgroup-agrawal-drift.

2.1 Subgroup Agrawal Drift Dataset

To explore the concept of localized drift, we define a synthetic dataset based on the Agrawal generator [1]. The Agrawal generator is commonly used for simulating data streams and generates samples x in a domain $\mathcal{D}$ with six numerical attributes and three categorical attributes, producing binary classification tasks. The attributes are as follows:

- salary, uniformly distributed from $20,000 to $150,000
- commission, 0 if salary has a value below $75,000, otherwise it is uniformly distributed from $10,000 to $75,000
- age, uniformly distributed from 20 to 80
- elevel (education level), uniformly chosen from 0 to 4
- car (car maker), uniformly chosen from 1 to 20
- zipcode (zip code of the town), uniformly chosen from 0 to 8
- hvalue (house value), uniformly distributed from $50,000 · zipcode to $100,000 · zipcode. Different zip codes, as such, are associated with different average house prices
- hyears (years the house has been owned), 1 to 30 uniformly distributed
- loan (total loan amount requested), uniformly distributed from $0 to $500,000

Each synthetic record is associated with a binary outcome (i.e., whether the loan is approved or not). Ten different functions, $f_0(x), f_1(x), \cdots, f_9(x)$ have been proposed in the original work to map a given record x to the binary ground truth value, $f_i : \mathcal{D} \to \{0, 1\}$. This completely defines any record $\{x, f_k(x)\}$ used either during training or in deployment. A perturbation can also be included so as not to make the classification task trivial. This perturbation (ranging from 0% to 100%) affects the attributes of x *after* the class has been assigned, thus adding an element of fuzzyness in the sample/class relationship.

A common technique to introduce concept drift [6] consists of adopting a classification function f_i for the original concept and a different one f_j ($i \neq j$) for the drift concept. At step t, the function is defined as a random variable F:

$$F = Z \cdot f_i(x) + (1 - Z) \cdot f_j(x), \qquad (1)$$

where $Z \sim Bernoulli(p_t)$. In this way, $F(x) = f_i(x)$ with probability p_t, $f_j(x)$ otherwise. The drift occurs gradually over time by defining p_t through a sigmoid function, $p_t = (1+e^{-4(t-k)/w})^{-1}$. k represents the center of the sigmoid function, and w is its width.

This drift, however, is applied uniformly to all samples. Instead, we aim to create a drift that is localized in nature, i.e., that only affects one subpopulation of the dataset. We consider a selector function $s : \mathcal{D} \rightarrow \{0, 1\}$, having value 1 for samples belonging to the target subgroup, 0 otherwise. We outline the definition of $s(\cdot)$ in more detail in Subsect. 2.2. We define F as follows:

$$F = s(x) \cdot [Z \cdot f_i(x) + (1-Z) \cdot f_j(x)] + [1 - s(x)]f_i(x). \tag{2}$$

In other words, the gradual drift is applied only to samples belonging to the target subgroup defined by $s(\cdot)$. All other samples will retain the original concept. We note that this definition can be extended to multiple subpopulations, which can be subject to different drifts.

2.2 Subgroup Definition

To simulate a localized drift, we need to define a target subgroup within the dataset. We produce meaningful subgroups by identifying slices of the domain, e.g., { age $\in$ [25, 30], salary $\in$ [\$75,000, \$100,000] }. We fully automate the synthetic dataset generation phase by introducing a random subgroup definition policy. This policy produces, for a desired subgroup size (i.e., subgroup support), a slice of the population that approximately encompasses it.

We adopt a greedy policy to identify a subset of slices that, combined, well approximate the target subgroup size. We do this by identifying random ranges of values (e.g., $[c, d]$) for randomly chosen attributes (e.g., attr $\sim U(a, b)$, $a \le c < d \le b$). The uniform distribution makes it trivial to compute the probability of belonging to the random range of values, as $P(\text{attr} \in [c, d]) = P(\text{attr}) = \frac{d-c}{b-a}$. Additionally, the attributes are independent from one another[1]. As such, their combined probability can be computed as the product of the separate probabilities, $P(\text{attr.1}) \cdot P(\text{attr.2}) \cdot \ldots \cdot P(\text{attr.n})$. We either include or discard a candidate slice based on whether it gets the current probability closer to the target one. Figure 2 provides an example where a subgroup of approximately the target size (10%) is iteratively defined by identifying a first slice on age, followed by a second one on salary. Because of the greedy nature of the algorithm, slices that do not provide an immediate improvement in terms of support are discarded. The algorithm terminates when either the subgroup size is within a tolerance threshold of the target one, or a maximum number of iterations is reached.

---

[1] As discussed in Subsect. 2.1, all but two attributes (commission and hvalue) are independently sampled from uniform distributions with known ranges. We only consider independent attributes for the definition of the target subgroup for simplicity.

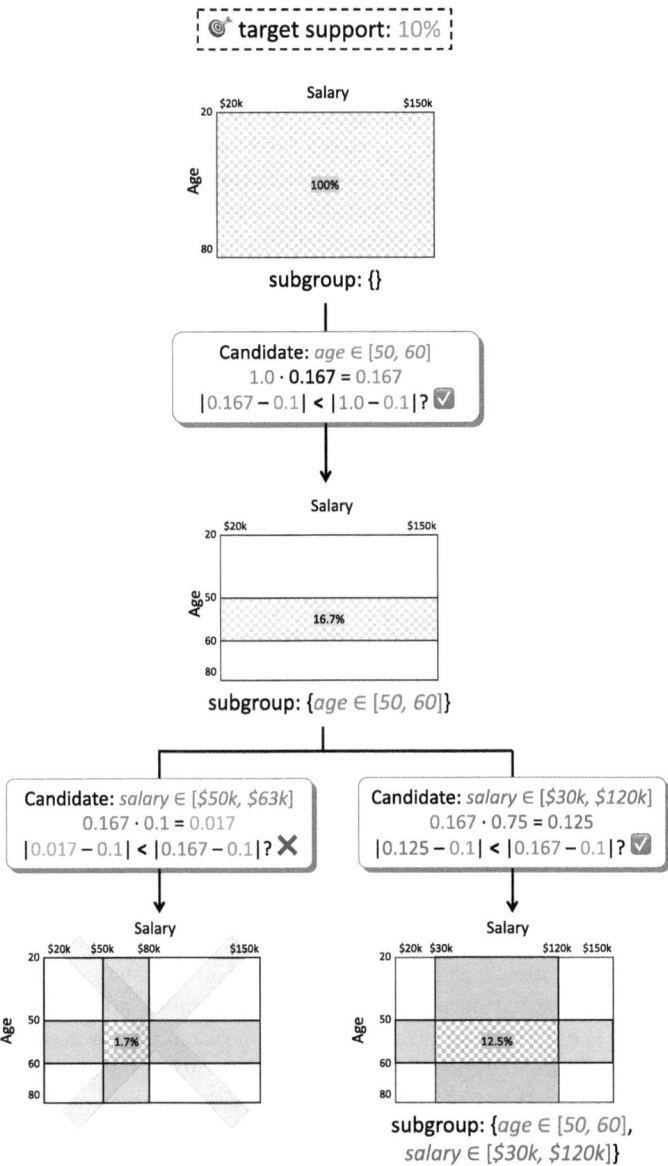

Fig. 2. Example of the greedy process adopted to randomly generate subgroups on 2 attributes. From top to bottom, the target subgroup is built by iteratively adding randomly generated slices of the attributes, if their inclusion produces a better approximation of the desired support (in the example, 10%). The checkered area at each step represents the size (support) of the current subgroup.

2.3 Examples of Generated Subgroups

In this subsection, we provide some instances of generated subgroups for various target subgroup sizes.

To guarantee variety in the generated subgroups, the proposed generator produces random subgroups that approximate the desired target size. As detailed above, the algorithm refines the generated subgroup until either the desired tolerance is reached or a maximum number of iterations has been executed. Figure 3 shows the distribution of gaps between target and obtained subgroup sizes, for a tolerance of 0.01 on the target size. In 83% of cases, the desired tolerance is reached, whereas in the remaining 17% of cases the maximum number of iterations (1,000) is reached.

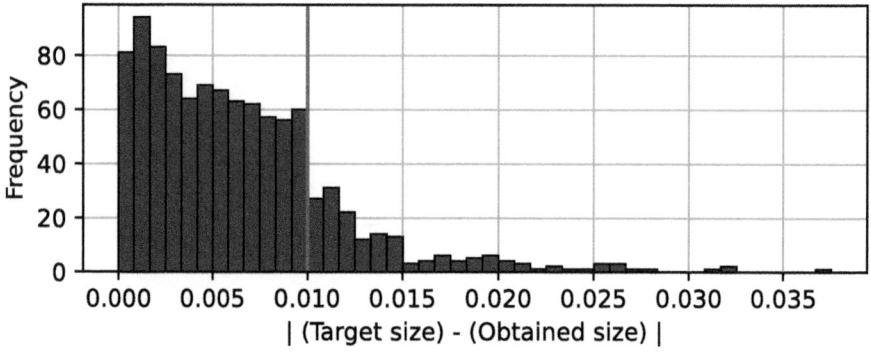

Fig. 3. Distribution of the absolute difference between target and corresponding obtained subgroup sizes, for 1,000 generated subgroups and various target sizes, tolerance of 0.01 (marked in red), maximum number of iterations set to 1,000. (Color figure online)

We report four examples of generated subgroups in Table 1. For each, we report the target (desired) size (5%, 10%, 25% and 50% of the population, respectively), the one computed according to the greedy policy adopted for subgroup generation, and the actual (empirical) size, as observed over a generated sample of 10,000 points. Both computed and actual sizes are close to the target one. If needed, the gap between computed and target subgroup sizes can be lowered by changing the maximum number of allowed iterations and/or the desired tolerance threshold.

3 Experimental Results

In this section, we show the performance of various drift detection techniques that are commonly adopted in literature on the proposed Subgroup Agrawal Drift dataset. We are mainly interested in the change in performance of these techniques as the size of the drifting subgroup changes.

Table 1. Examples of subgroups generated for various target sizes. The *computed size* reports the expected subgroup size, and the *actual size* represents the measured subgroup size for a sample of 10,000 instances.

Generated subgroup	Target size	Computed size	Actual size
{ $elevel \in [0,3) \wedge$ $zipcode \in [6,7) \wedge$ $age \in [29,78)$ }	0.05	0.0536	0.0552
{ $car \in [15,19) \wedge$ $salary \in [39000, 116000) \wedge$ $zipcode \in [0,8)$ }	0.1	0.1045	0.107
{ $zipcode \in [2,5) \wedge$ $salary \in [30000, 139000) \wedge$ $age \in [22,80) \wedge$ $car \in [1,20)$ }	0.25	0.2505	0.2527
{ $elevel \in [1,4) \wedge$ $age \in [20,78) \wedge$ $salary \in [21000, 140000) \wedge$ $hyears \in [1,30)$ }	0.5	0.501	0.4965

3.1 Drift Detection Techniques

We considered the following drift detectors.

- **DDM (Drift Detection Method)** [4] is a statistical technique that monitors the error rate of a model over time. When the error rate increases significantly, it indicates a possible change in the data distribution. If this increase surpasses a pre-defined drift threshold, DDM triggers the detection of a drift.
- **EDDM (Early Drift Detection Method)** [2] improves upon DDM by focusing on the distance between errors instead of just the error rate. This method aims to detect gradual changes more effectively. It calculates the average distance between errors and monitors the standard deviation of these distances. Significant changes in these metrics can indicate a drift, allowing the model to adapt more quickly to evolving data streams.
- **HDDM (Hoeffding Drift Detection Method)** [3] is based on Hoeffding's inequality, which provides a way to determine the bounds of an estimator with high probability. HDDM uses this statistical method to detect changes in the distribution of incoming data compared to older data. By comparing the distributions of recent data to older data, HDDM can identify when a significant change has occurred, suggesting that the underlying data distribution has drifted.
- **FHDDM (Fast Hoeffding Drift Detection Method)** [7] is an enhanced version of HDDM, designed to provide faster and more accurate detection. It applies Hoeffding's bounds to smaller windows of data, allowing it to detect drifts more quickly and with fewer false alarms. FHDDM is particularly useful in scenarios requiring rapid adaptation to changing data.

For each method, we identify the best-performing configuration of hyperparameters through a grid search on the dataset.

3.2 Dataset

We adopt the proposed synthetic dataset for benchmarking drift detection techniques as the drifting subgroup sizes vary. In particular, we are interested in the performance when the drifting subgroups are small, as these are the drifts that are intuitively more likely to go undetected. We sample subgroup sizes from 1% to 100% (i.e., the full population) logarithmically.

For each subgroup size, we conduct 100 experiments. For half of them, we inject drift to a random subgroup of the desired size (positive experiments). The other half is instead not injected with any drift (negative experiments).

For positive experiments, we randomly choose one out of the 10 classification functions for the original concept, and a different one for the drift concept. For negative experiments, we instead use a single concept throughout the entire experiment. For all experiments, we build a training set comprised of 10,000 points sampled from the underlying distribution and associated with the original concept. We train a simple decision tree classification model with a depth of up to 5 nodes on this training set. Subsequently, we sample 200 batches of data (1,000 points each). For positive experiments, the concept drift is injected gradually, as detailed in Subsect. 2.1. The injection is centered around the 100th batch, with a width of 100 batches. The subgroup accuracy in Fig. 1 provides a visual intuition of the setting. We introduce a perturbation of 25% of the input attributes to make the classification problem non-trivial. For each experiment, the various drift detection techniques are used to monitor and potentially detect drifts. Since each algorithm can potentially produce multiple drift detections, we count the number of detections. We determine the threshold on the minimum number of detections to trigger a drift alert using a ROC curve computed on 30% of the experiments. We use the rest of the experiments to compute the performance in terms of accuracy, F_1 score, False Positive Rate (FPR) and False Negative Rate (FNR), of various drift detection techniques.

3.3 Results

Figure. 4 summarizes the main results. Both accuracy and F_1 highlight how all considered techniques achieve near-perfect performance in detecting drifts when the drifting subgroup is large enough (approximately 10% of the dataset or more). Instead, none of the approaches achieved satisfactory results for lower support sizes. To better understand the cause of this drop in performance, we additionally computed the FPR and FNR for each technique for various sizes of drifting subgroups.

Interestingly, the FPR is largely unaffected by the size of the drifting subgroup. In other words, none of the considered approaches produces an excess of false positive predictions when smaller subgroups are drifting. This is in accordance with what was expected: drifts of smaller subgroups go unnoticed, meaning that fewer positive predictions are produced overall.

Instead, the FNR plot presents a different perspective. In this case, it is clear that there exists an abundance of false negatives when the drifting subgroups are smaller in size. These false negatives are drifts that are not being detected: as expected, the various drift detection techniques cannot handle properly drifts of smaller subpopulations.

As the drifting subpopulations grow, the number of false positives produced decreases. Some approaches, such as DDM, have an earlier and sharper reduction in FNR, whereas other approaches (more significantly, EDDM) have a delayed response, meaning that they struggle to detect drifts even when the target subgroups are larger.

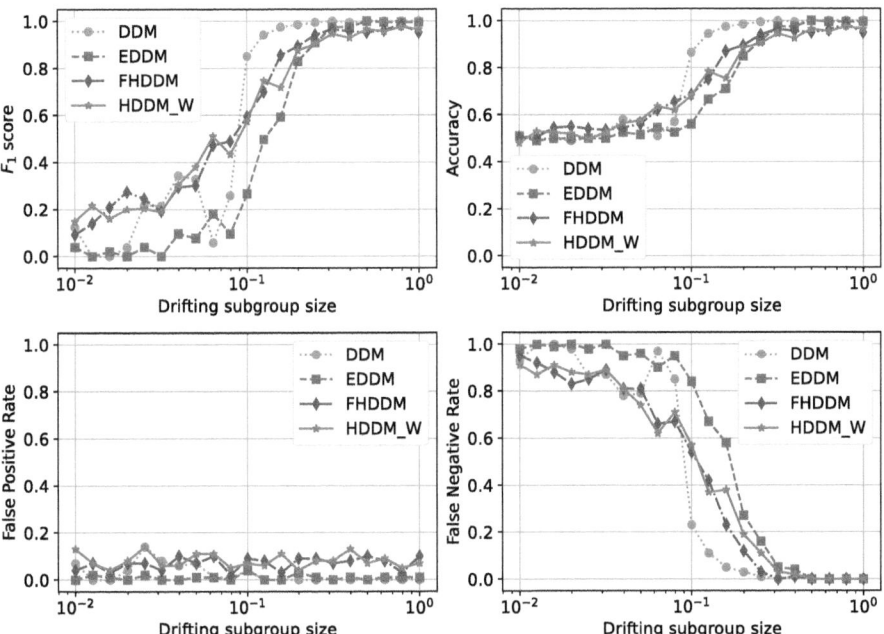

Fig. 4. Performance, in terms of F_1 score, accuracy, False Positive Rate, and False Negative Rate, of various drift detection techniques on the binary task of detecting the occurrence of drift events. Subgroup Agrawal Drift dataset, perturbation of 25%, various sizes of drifting subgroups.

4 Conclusions

In this work, we highlighted a problem that affects commonly adopted drift detection techniques: drifts are only detected if they affect a large fraction of the original data. This implies that drifts affecting smaller subpopulations (e.g., minorities) may go undetected. This is problematic, since it implies that models may be silently drifting and underperforming for certain populations. To benchmark the performance of various detectors under subgroup drifts, we introduce the Subgroup Agrawal Drift Dataset, a synthetic data generator that injects a specific subgroup of a desired size with noise. The experimental results show indeed that commonly adopted techniques only detect subgroup drifts when these cover a large fraction of the dataset, producing a large number of false negatives in the case of smaller diverging subgroups. As a natural next step, we plan on addressing this shortcoming of current drift detection techniques. A first step in this direction has been taken by proposing a subgroup-aware drift detection technique [5].

Acknowledgements. This work is partially supported by the FAIR - Future Artificial Intelligence Research (PIANO NAZIONALE DI RIPRESA E RESILIENZA (PNRR) - MISSIONE 4 COMPONENTE 2, INVESTIMENTO 1.3 - D.D. 1555 11/10/2022, PE00000013) and the spoke "FutureHPC & BigData" of the ICSC - Centro Nazionale di Ricerca in High-Performance Computing, Big Data and Quantum Computing, both funded by the European Union - NextGenerationEU. This manuscript reflects only the authors' views and opinions, neither the European Union nor the European Commission can be considered responsible for them.

References

1. Agrawal, R., Imielinski, T., Swami, A.: Database mining: a performance perspective. IEEE Trans. Knowl. Data Eng. **5**(6), 914–925 (1993)
2. Baena-Garcıa, M., del Campo-Ávila, J., Fidalgo, R., Bifet, A., Gavalda, R., Morales-Bueno, R.: Early drift detection method. In: Fourth International Workshop on Knowledge Discovery from Data Streams, vol. 6, pp. 77–86 (2006)
3. Frias-Blanco, I., del Campo-Ávila, J., Ramos-Jimenez, G., Morales-Bueno, R., Ortiz-Diaz, A., Caballero-Mota, Y.: Online and non-parametric drift detection methods based on hoeffding's bounds. IEEE Trans. Knowl. Data Eng. **27**(3), 810–823 (2014)
4. Gama, J., Medas, P., Castillo, G., Rodrigues, P.: Learning with drift detection. In: Bazzan, A.L.C., Labidi, S. (eds.) SBIA 2004. LNCS (LNAI), vol. 3171, pp. 286–295. Springer, Heidelberg (2004). https://doi.org/10.1007/978-3-540-28645-5_29
5. Giobergia, F., Pastor, E., de Alfaro, L., Baralis, E.: Detecting interpretable subgroup drifts. arXiv preprint arXiv:2408.14682 (2024)
6. Montiel, J., et al.: River: machine learning for streaming data in Python (2021)
7. Pesaranghader, A., Viktor, H.L.: Fast Hoeffding drift detection method for evolving data streams. In: Frasconi, P., Landwehr, N., Manco, G., Vreeken, J. (eds.) ECML PKDD 2016. LNCS (LNAI), vol. 9852, pp. 96–111. Springer, Cham (2016). https://doi.org/10.1007/978-3-319-46227-1_7

Unsupervised Concept Drift Detection Based on Parallel Activations of Neural Network

Joanna Komorniczak[(✉)] and Paweł Ksieniewicz

Department of Systems and Computer Networks, Wrocław University of Science and Technology, Wrocław, Poland
joanna.komorniczak@pwr.edu.pl

Abstract. Practical applications of artificial intelligence increasingly often have to deal with the streaming properties of real data, which, considering the time factor, are subject to phenomena such as periodicity and more or less chaotic degeneration – resulting directly in the *concept drifts*. The modern concept drift detectors almost always assume immediate access to labels, which, due to their cost, limited availability, and possible delay, has been shown to be unrealistic. This work proposes an unsupervised *Parallel Activations Drift Detector*, utilizing the outputs of an untrained neural network, presenting its key design elements, intuitions about processing properties, and a pool of computer experiments demonstrating its competitiveness with state-of-the-art methods.

Keywords: concept drift · data streams · unsupervised drift detection · neural networks

1 Introduction

The modern world is dominated by the mass production of data transmitted daily in petabytes of information traveling over the internet [9]. Artificial intelligence applications attempt to organize this chaotic reality from the level of dispersed information into knowledge, recently most often relying on *semi-supervised* and *unsupervised learning* methods, significantly reducing the need to rely on human experts in the labeling process [6]. However, default solutions of this type treat available data as static in time, often ignoring the phenomena of knowledge historicity and periodicity of concepts, thus striving to maximize efficiency within the immense volume of *Big Data* [8].

The field of *machine learning* that focuses more on data velocity and considers the possibility of changing concepts between successive batches of incremental processing is *Data Stream Processing*. One of the critical issues in this field is *drift detection*, which involves identifying solutions that allow for effective signaling of significant changes in the concept. It should be noted, however, that the most common state-of-the-art drift detectors most often assume full labeling of the data stream, which does not fit well into the increasingly dominant paradigm of *semi-supervised* and *unsupervised learning*. This shows the significant need to develop research on unsupervised drift detectors, potentially enabling broad applications of *Data Stream Processing* achievements in mainstream artificial intelligence research.

Concept Drift Phenomenon. Concept drift is taxonomically divided in terms of three main axes [1]. According to the impact on recognition ability, drifts are divided into *real* ones, the influence of which is visible when monitoring the quality of classification, and *virtual*, which do not affect the decision boundary but may constitute the initial stage of *real* changes. According to dynamics, distribution shift may occur at a single point in time (*sudden* drift), or the transition can be spread over a longer period in *gradual* and *incremental* drifts. During *incremental* drift, a temporary concept between the initial and target ones is observed, while in the case of *gradual* change, objects from two consecutive concepts co-occur during the transition period. Finally, according to drift *recurrence* – a concept from the past may recur due to cyclical phenomena such as seasons or daily cycles. Additionally, the taxonomy considers drifts in which *prior probability* of the classification problem changes [19]. Such drifts may affect recognition quality, showing falsely high accuracy values or a decrease in the quality when using metrics dedicated to imbalanced data, without a drift directly affecting the decision boundary.

2 Related Works

According to the guidelines described by Domingos et al. [7], a critical element of processing data streams is the mechanism for adapting to concept drifts. In the face of such changes in the data stream, two approaches are used: *continuous rebuild* and *triggered rebuild* [22]. In the case of the continuous rebuild, classifiers are trained throughout the entire processing period. In contrast, in the case of the triggered rebuild, specific determinants are used to indicate drifts, and only after the change is detected, the classifiers are updated to the current state of the posterior distribution of a stream. All approaches from the continuous rebuild strategy require almost immediate access to labels, which are used to incrementally train the classifier. The factors used in the *triggered rebuild* approach can be further divided into three categories: (*a*) those monitoring the classification model, (*b*) those monitoring the data, and (*c*) those monitoring the output from the classification model [17]. Similarly to the *continuous rebuild*, the methods monitoring the classification output use labels for recognition quality assessment.

In the *triggered rebuild* approach, the concept drift detectors are responsible for signaling the need to update the model. The first proposed drift detection methods took advantage of the fact that *real* concept changes affect the recognition quality and monitored the frequency or distance of errors made by the classifier. Examples of such methods are *Drift Detection Method* (DDM) [10] and *Early Drift Detection Method* (EDDM) [3]. Subsequent detection methods used more complex mechanisms based on sliding windows in the *Adaptive Windowing* (ADWIN) [5] algorithm, pairs of classifiers in *Paired Learners* [2], and ensembles of classifiers in *Diversity for Dealing with Drifts* approach [26]. The main disadvantage of these solutions is a strong dependence on label access. It is worth mentioning here that *implicit* supervised detectors have also been proposed, which do not directly rely on the classification quality to detect concept changes but use labels to analyze algorithm-independent properties of the data [15]. Regardless of how labels are used, the assumption of their almost immediate availability is not realistic due to limited access [29], their cost [24], and possible time delay [14]. For those reasons, scientific interest in unsupervised drift recognition methods has increased in recent years [11]. While some unsupervised methods monitoring the classification model will require access to labels, this is mainly to update the model after change detection [29].

Following the taxonomy described earlier, unsupervised drift detection methods will use two types of factors: those dependent on the classification model and those dependent on the data distribution itself. The data distribution is monitored in the *Nearest Neighbor-based Density Variation Identification* (NN-DVI) [23] detector using the *k-nearest neighbors* algorithm. Similarly, grid-based data distribution monitoring is proposed in the *Grid Density based Clustering* (GC3) [30] approach. In the *Centroid Distance Drift Detector* (CDDD) [16] method, the distance between the centroids of subsequent batches is examined. This method can operate in both supervised and unsupervised modes. There are also methods based on the analysis of outlier observations, such as *Fast and Accurate Anomaly Detection* (FAAD) [21], proposed mainly for the purpose of anomaly detection. In *One-Class Drift Detector* (OCDD) [13], a one-class classifier is used to examine the percentage of objects recognized as not belonging to the recognized concept. A similar strategy was used in *Discriminative Drift Detector* (D3) [12], where a discriminative classifier is used instead of a one-class classifier to explicitly recognize objects from the new concept from those from the previous one in variable-width windows.

Among the solutions based on the classifier properties, the *Margin Density Drift Detection* (MD3) [29] method should be mentioned, in which the density of samples near the decision boundary of the SVM classifier is examined. Similarly, in the *Confidence Distribution Batch Detection* (CDBD) [22] algorithm, the confidence of a classifier is monitored. Both of those approaches, despite their unsupervised detection, require access to labels in order to rebuild the monitored classifier in the case of a drift.

Motivation and Contribution. In this work, we present a fully unsupervised *Parallel Activations Drift Detector* (PADD) method interpreting the activations of randomly initialized neural network (NN). Its overall detection mechanism shows some similarities to the CDBD detector – which uses confidence in the outputs from the trained classifier (possibly NN) – but without the requirement of label access to update the model in the event of drift. Similarly to GC3 method, PADD employs the paradigm of original sample transformation into the condensed space, but on the contrary, it does not use a regular distribution grid, introducing non-uniform, tangled set of projections typical for the initial random state of a NN [28].

The main contribution of this publication is a new drift detection method, operating on a fully unsupervised analysis of raw NN activations. The work validates the overall detection quality on the synthetic data streams with various characteristics, comparing the proposed approach with unsupervised and supervised state-of-the-art drift detection methods. Conducted experiments are publicly available in a *GitHub* repository[1] to preserve the replicability of the research.

3 Parallel Activations Drift Detector

This work presents an unsupervised drift detection method, operating purely on the output of a randomly initialized NN. The approach is described in the Algorithm 1.

Algorithm 1. Pseudocode of the *Parallel-Activations Drift Detection*

Input:
  $\mathcal{DS} = \{\mathcal{DS}_1, \mathcal{DS}_2, \ldots, \mathcal{DS}_k\}$ – data stream,
Symbols:
  α – significance level for statistical test
  θ – threshold for drift detection
  r – statistical test replications performed for each NN output
  s – number of samples drawn for statistical test
  $\mathcal{C}$ – stored activation values for all NN outputs since last drift
  $\mathcal{NN}()$ – forward pass from NN model with random weights and e outputs
  $\mathcal{S}()$ – statistical test for sample subset comparison

1: **for all** $\mathcal{DS}_k \in \mathcal{DS}$ **do**
2:    $c \leftarrow \mathcal{NN}(\mathcal{DS}_k)$ ▷ Get ultimate network activations for current data chunk
3:    **if** $\mathcal{C}$ is not empty **then**
4:      $a \leftarrow 0$ ▷ Initialize counter as zero
5:      **for all** $e_i \in e$ **do**
6:        **for all** $r_i \in r$ **do**
7:          $cc \leftarrow$ random s from $c[e_i]$ ▷ Randomly select samples from current and past certainty
8:          $pc \leftarrow$ random s from $\mathcal{C}[e_i]$
9:          $p \leftarrow \mathcal{S}(pc, cc)$ ▷ Compute p-value of statistical test
10:          **if** $p < \alpha$ **then**
11:            increment a
12:          **end if**
13:        **end for**
14:      **end for**
15:      **if** $a > \theta \times e \times r$ **then**   ▷ Drift detected
16:        $\mathcal{C} \leftarrow \emptyset$
17:        indicate drift in chunk k
18:      **end if**
19:    **end if**
20:    store c in $\mathcal{C}$ ▷ Store current activations for future comparison
21: **end for**

The method processes data streams divided into non-interlacing batches ($\mathcal{DS}_k \in \mathcal{DS}$). Drift detection is marked based on r replications of statistical tests

---

[1] https://github.com/w4k2/padd.

– aiming to validate the null hypothesis stating the lack of significant difference between two groups of independent measurements – comparing (*a*) a sample of size *s* from the distribution of past and (*b*) current activations at the all *e* outputs of the NN. The initial – and constant during the full processing – random weights of NN are drawn from a normal distribution. As default settings, the normal distribution has an expected value of zero and a standard deviation of 0.1. The statistical test for distribution comparison is the Student's T-test for independent samples.

The critical parameters of the method are the significance level *alpha* (α) of the statistical test and the *threshold* parameter (θ), indicating the fraction of all tests that need to signal statistical independence of distributions to induce a drift detection.

At the beginning of the processing of each batch, *c* activations are calculated for samples from a given batch at all NN outputs. In the first chunk, the historical activations $\mathcal{C}$ are yet unknown, so the detection step is skipped due to the lack of reference data. The current activations *c* are stored in the pool of historical outputs ($\mathcal{C}$). Otherwise, statistical tests are performed for individual network outputs. In lines 5:14 of the pseudocode, *r* replications of the statistical test are performed for each e_i output of the network. Samples of size *s* are drawn with replacement from historical activations for a given output $\mathcal{C}$ and for the current distribution *c*. If the statistical test shows a significant difference between the past and current distribution, the counter *a* is incremented. The detection criterion is described in lines 15:18 of the pseudocode. If the counter *a* exceeds the required number of tests showing statistically significant (difference defined using θ, the number of outputs *e*, and the number of test replications *r*) a concept drift in the current chunk is signaled. Such a signal implicates clearing the buffer of past supports ($\mathcal{C}$). For each batch, the current activations *c* are saved to the historical data at the end of processing (line 20).

The Student's T-test shows a noticeably high sensitivity to the sample selection from the random variable provided to it. Therefore, the PADD method stabilizes its verdict with replication of the measurement, which is possible thanks to a reliable buffer of historical activations. The invariance of the model weights, in turn, preserves the repeatability of the transformations performed by the NN, which should lead to results of low-dimensional embeddings to be statistically dependent in the absence of changes in the posterior distribution of the stream – which we associate with both real and virtual drift phenomenon. Consequently, the proposed method is not built around the observation of the decision boundary – as is the case with solutions basing detection on the evaluation of significant changes in the quality of processing – but presents the potential to register general changes in the distribution occurring regardless of a given label bias.

Figure 1 shows the intuition behind the method operation on the exemplary stream with 250 data chunks and a final NN layer with four outputs. The first line presents the image (probing of a model with a mesh-grid covering two-dimensional feature space) of the NN output in the area sampled by data distribution. Red regions correspond to high activations, and blue to low activations.

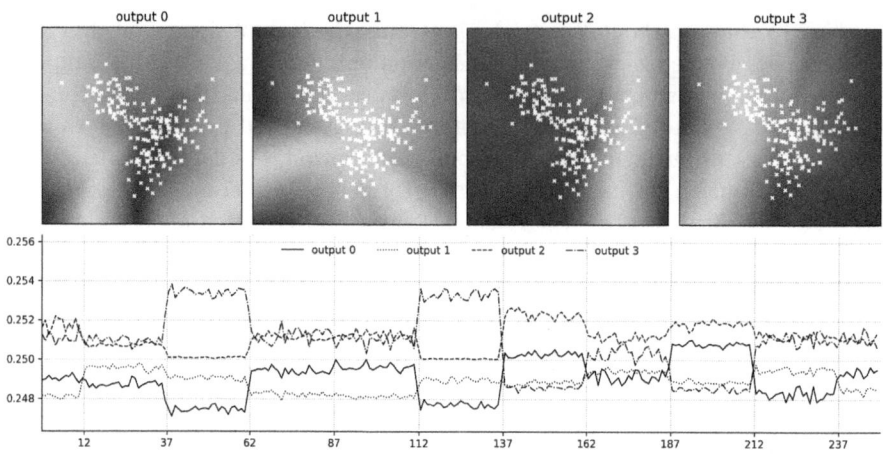

Fig. 1. An example of 2-dimensional data (white points) presented in a context of ultimate layer activations of randomly initialized NN (top charts) and their mean activation (bottom chart) of four examined NN outputs during stream processing. Vibrant red areas in the top charts correspond to the high activation of a model, and vibrant blue to its low activation. The ticks on the horizontal axis of the bottom chart signal the moments of an abrupt concept change. (Color figure online)

Random initialization of the weights causes diversified local landscapes in every dimension of the output space. In these areas, samples from one batch are marked with white markers. If drift occurs and the posterior distribution of a sampled data chunk changes, the structure of pseudo-supports in the recognized set will map this change within all or a part of the NN outputs. The second row of the figure presents the average activation values for batches in the data stream across all four outputs. The moments when drifts occur are clearly visible, additionally marked with chunk indices on the horizontal axis of the chart. In relation to the state of the network, some drifts will be easier to identify than others – for example, the difference between the average output for the first drift (chunk 12) is less visible than for the next drift (chunk 37).

4 Experiment Design

The experiment and methods were implemented in Python. The implementation of the proposed method, the experimental code and the results are publicly available.

4.1 Data Stream Generation

The experimental evaluation was performed on synthetic streams obtained using the generator from the *stream-learn* library [20]. The streams were described with various number of features and the number and type of drifts. The streams were processed in 250 batches of 200 samples, and each stream with a specific configuration was generated ten times with varying random states of a generator.

The complete configuration of data streams is presented in Table 1. The experiment compared the methods on the complete pool of 240 data streams.

Table 1. Generator configuration for synthetic data streams

PARAMETER	CONFIGURATION	PARAMETER	CONFIGURATION
Number of chunks	250	Drift frequency	3, 5, 10, 15 drifts
Chunk size	200	Number of features	30, 60, 90 (30% informative)
Drift dynamics	sudden, gradual	Replications	10

The choice of synthetic data was dictated by the option to verify the operation of detection methods in various conditions and the possibility of replicating the stream generation to stabilize the results for statistical analysis. Additionally, only in the case of synthetic data the exact moments of drifts are known [25]. This enables comparing changes signaled by the methods with the actual concept drifts.

4.2 Measuring Quality of Drift Detection

Three *drift detection error* measures were used to assess the methods' detection quality [18], evaluating the similarity of drifts occurring in the stream to those detected by a given method. The comparison protocol based on the classification quality has proven to show no relationship between the detection quality and the accuracy of classification [4].

The three *drift detection error* measures will evaluate detectors on three different criteria:

D1 – *The average distance of each detection to the nearest drift*
This measure will penalize the methods with numerous redundant drift detections, occurring far from the moments of actual concept changes. The average distance from detection to drift will increase in the case of hypersensitive methods. However, this measure does not consider the unrecognized concept changes – hence, methods of low sensitivity will not be penalized for lack of specific drift detection.

D2 – *The average distance of each drift to the nearest detection*
This measure considers the closest detection of each actually occurring drift. In this measure, conversely to $D1$, the methods will be penalized for lack of drift signalization. However, the hypersensitivity of drift detectors will have no impact on this error measure – as only the single detection closest to drift will be considered.

R – *The adjusted ratio of the number of drifts to the number of detections*

This error measure considers the number of drifts that actually occur in the stream and the number of detections signaled by the method. The lowest possible error value of zero will be achieved for a method signaling the exact number of changes as a number of drifts. In case the method signals multiple redundant detections, the error will rise towards the value of one. If the method signals fewer detections due to low sensitivity, the error can rise to values above one. This measure, therefore, penalizes both too many and too few detections. Nevertheless, too little detection brings the error value higher, assuming that the lack of drift recognition is of greater significance during stream processing.

It is essential to note that measures can only be defined if the evaluated method signals any detection. Otherwise, the errors will be infinite, and statistical comparison will not be possible.

4.3 Goals of the Experiments

Hyperparameter Selection. The first experiment aimed to select appropriate hyperparameters of the proposed method. Out of the five available hyperparameters, only the two most critical were optimized: *alpha* and *threshold*. The remaining ones were fixed:

- the number of network outputs e was 12,
- the number of statistical test replications r was 12,
- the sample size s was 50.

A neural network with a single hidden layer containing ten neurons and a *ReLU* activation function was used.

For the *alpha* parameter, 15 values evenly sampled from the range from 0.03 to 0.2 were tested, and for the *threshold* parameter, we evaluated ten equidistant values from the range 0.1 to 0.3. The operation of the method with the indicated configurations was tested for the streams with ten drifts. The result of this experiment will indicate the range of values of these two parameters for which the method effectively recognizes drifts. Since both examined parameters indicate sensitivity to changes, their relationship should be visible.

Comparison With the Reference Methods. The second experiment aimed to compare the proposed approach with reference methods. State-of-the-art supervised and unsupervised detectors were selected. The research was carried out with the methods' implementation provided by authors, or, when the method enforced the *online* analysis of samples, the provided implementation was modified to allow processing streams in the form of batches.

Table 2 presents all methods considered in the experiment, including the proposed PADD method. The first column shows the acronym of the method, the second the category in the context of label access, the third the full name of the method, and a reference to the article introducing this approach. The last column describes the hyperparameterization of the method used in the experiment.

Table 2. Method configuration for experimental evaluation

ACRONYM	CATEGORY	METHOD NAME	SELECTED HYPERPARAMETERS
MD3	Unsupervised with label request	Margin Density Drift Detection [29]	*threshold* parameter set depending on number of features: 0.15 for 30 features, 0.1 for 60 features and 0.08 for 90 features
OCDD	Unsupervised	One-Class Drift Detector [13]	*percentage* parameter set depending on problem dimensionality: 0.75 for streams with 30 features, 0.9 for 60 features and 0.999 in case of 90 features
CDDD	Unsupervised	Centroid Distance Drift Detector [16]	*sensitivity* parameter set depending on concept drift density: 0.2 for streams with sparse changes (3,5) and 0.9 for streams with dense changes (10,15)
PADD	Unsupervised	Parallel Activations Drift Detector	configuration based on the results from the first experiment: *alpha* equal to 0.13 for gradual drift and 0.07 for sudden; *threshold* of 0.26 for gradual drift and 0.19 for sudden; $r = 12$; $e = 12$; $s = 50$
ADWIN	Supervised	Adaptive Windowing [5]	default *delta* of 0.002, the base classifier used for error monitoring was Gaussian Naive Bayes
DDM	Supervised	Drift Detection Method [10]	default detection *threshold* of 3, the base classifier used for error monitoring was Gaussian Naive Bayes
EDDM	Supervised	Early Drift Detection Method [3]	default *beta* of 0.9, the base classifier used for error monitoring was Gaussian Naive Bayes

The default parameters were selected for supervised methods, consistent with the implementation in the *scikit-multiflow* library [27]. For the remaining approaches, the hyperparameters were manually selected according to the articles introducing the methods or altered to effectively process the evaluated types of streams.

5 Experimental Evaluation

5.1 Hyperparameter Selection

The first experiment aimed to select the appropriate method hyperparameters, focusing on two variables describing the method sensitivity – *alpha* value for statistical test significance and *threshold* value for method integration.

The results for three drift detection error measures and a single stream described by 30 features and characterized by sudden concept drifts are presented in Fig. 2. The heatmaps present the $D1$, $D2$ and R errors, respectively. Blue cells indicate low error values, while red ones indicate high errors. The cells that are left blank indicate that the error was not possible to be measured due to the lack of detection. Each heatmap's vertical axis describes the *alpha* parameter values, and the horizontal axis describes the *threshold* parameter.

The lower left area of each heatmap – for the low *threshold* values and high *alpha* – describes a high sensitivity to changes present in the stream and is related to numerous redundant detections, visible in high $D1$ error and significant R error. Meanwhile, the upper right area describes the low sensitivity of a method

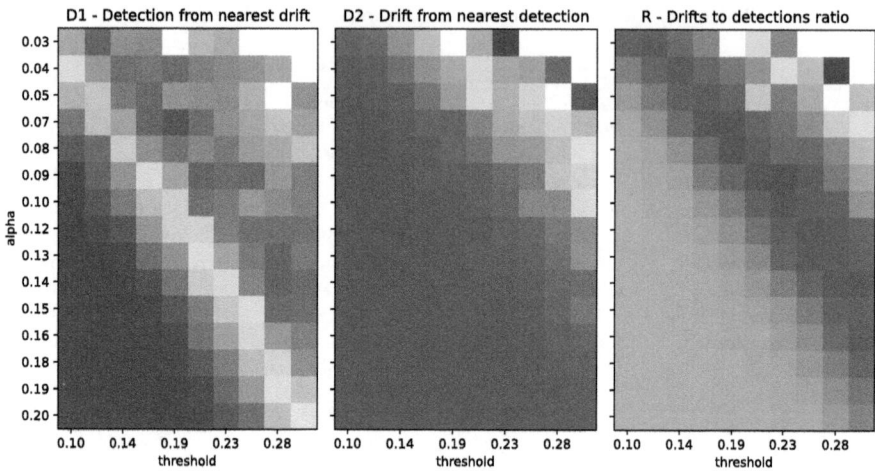

Fig. 2. Drift detection error measures for a single 30-dimensional data stream with sudden concept drifts, depending on the values of two critical hyperparameters – *alpha* and *threshold*, describing the method sensitivity. Red cells indicate high errors, while blue cells – the low error. (Color figure online)

– where all error values, including $D2$, are high. The white cells, visible for the low sensitivity parameter values, result from a lack of returned detections and an inability to calculate the drift detection error values. The best configuration of the method will be parameter values between these two extremes, where all three measures return low error.

Figure 3 presents the results of the experiment for all streams analyzed in the first experiment. The color heatmaps show the combination of three *drift detection error* measures after their normalization. The red color channel corresponds to $D1$ measure, green to $D2$, and blue to R error. The results for streams with sudden drifts are presented in the first row, and for gradual concept change, they are presented in the second. The columns present various dimensionalities of the data – 30, 60, and 90 features, respectively. After such a color combination of error values, the lowest errors in all three criteria will be marked by colors close to black.

It is worth noting that the method's effectiveness is highly dependent on the selection of these two parameters, and the lack of precise selection may result in excessive detection or failure to recognize drifts. Ultimately, the following parameter combinations were selected for the comparison experiment: for *gradual* drifts – *alpha* equal to 0.13 and *threshold* equal to 0.26; for *sudden* drifts – *alpha* equal to 0.07 and *threshold* of 0.19.

5.2 Comparison with Reference Methods

The results for streams with ten drifts are shown in Fig. 4. The columns indicate the results for different numbers of features – from 30 in the first column to 90 in

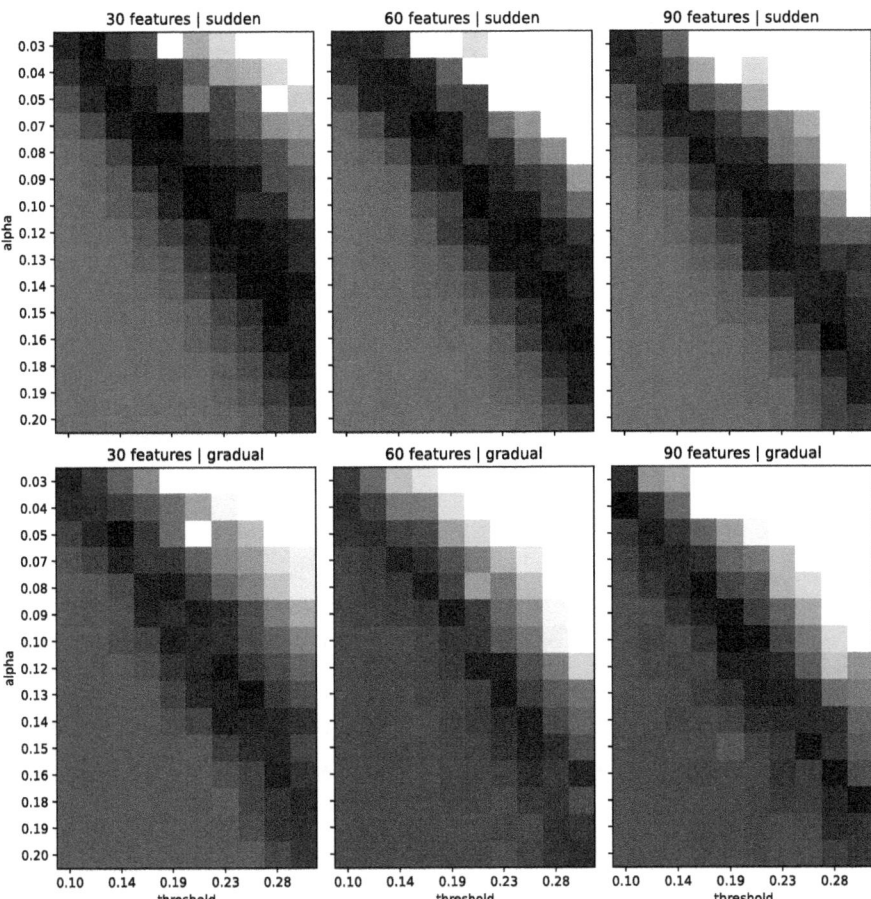

Fig. 3. The combination of drift detection error measures, normalized to a range 0–1 in each of three measures, presented as an RGB image. Black cells indicate the lowest error across all three measures. (Color figure online)

the last one – and the rows for different numbers of drifts present in the stream – from 3 drifts in the first row to 15 drifts in the last row. On the horizontal axis of each graph, successive chunks of the data stream are visible, while the central moments of the actual drift are marked with ticks on the x-axis and a grid. Each detection is marked with a single point. For emphasis, the proposed approach is shown in red. The consecutive lines show the results from subsequent replications for a given detector. The last row shows the concept drift dynamics.

These plots allow to visually assess the detection quality. The ideal result would be for the method to flag a single change at the center point of each drift occurrence across all replications – this would result in vertical lines overlapping with the drift markings on the horizontal axis.

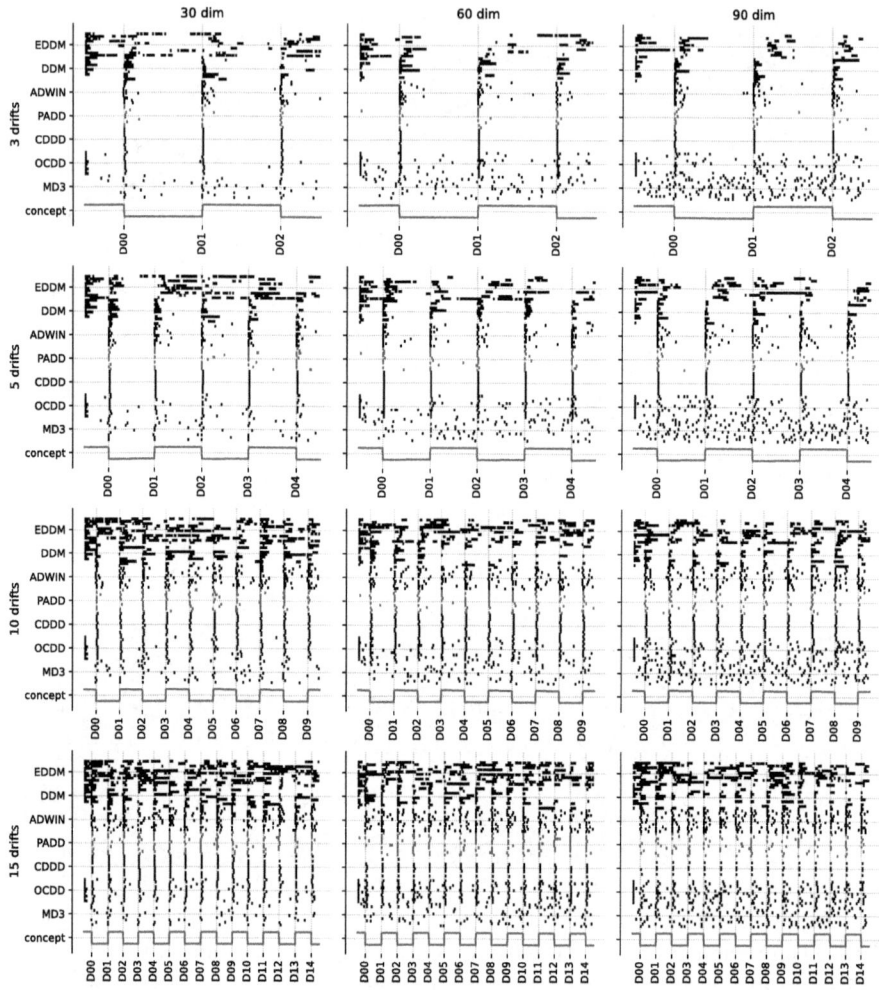

Fig. 4. Detection moments of all evaluated methods for streams with *sudden* concept drifts. The actual drifts are marked with ticks on the horizontal axis and particular methods's detections in ten stream replications with black or red (in the case of the proposed approach) points. The ideal drift detection would result in vertical lines overlapping with ticks on x-axis. (Color figure online)

Figure 5 shows the results for streams with gradual concept drift similarly. Notice that the detections for all methods are more dispersed in the case of gradual drifts – this is either due to the detection in an early or later phase of drift or due to multiple signaling of the same drift stretched over time in the case of high sensitivity of the method.

Looking at the color-coded tables in the form of heat maps, shown in Fig. 6, allows for a more general interpretation. The low error values are marked with

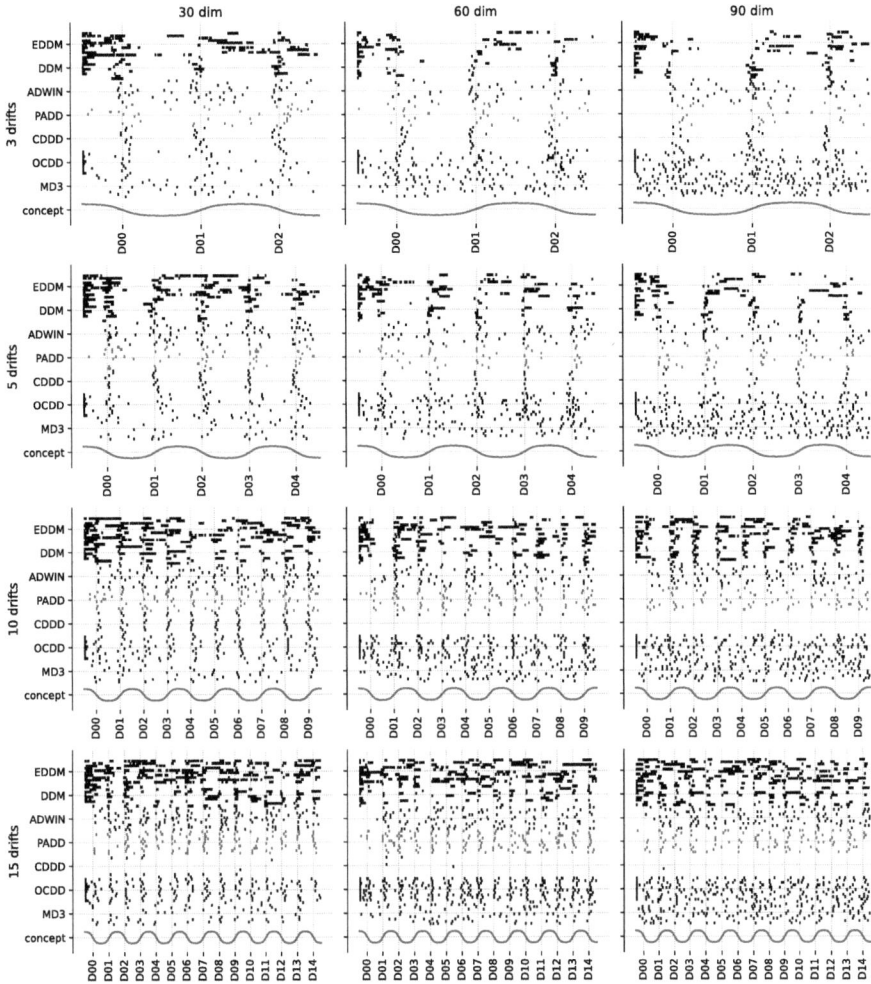

Fig. 5. Detection moments of all evaluated methods for streams with *gradual* concept drifts. The ticks on x-axis indicate the central point of each concept drift.

blue, and the high ones with red. Cases for which it was impossible to calculate errors – due to lack of detection – are left as blank cells. The columns of the figure show the results for the three drift detection error measures – $D1$, $D2$, and R, respectively. In each heat map, the results for individual methods are presented in columns, and the results for all tested types of streams are presented in rows, averaged over ten replications.

In the first error, $D1$, the average distance to the nearest drift is measured for each detection. This means that methods that signal many distant drifts from the moment of their actual occurrence, such as DDM, EDDM, and MD3, will achieve higher error values. It is also visible that for streams where drifts are

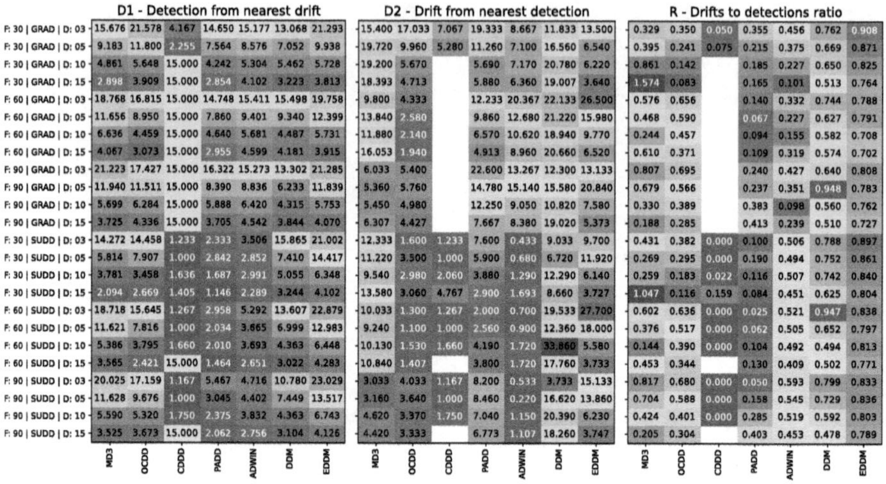

Fig. 6. Drift detection error measures: D1 (left), D2 (center), and R (right) for all evaluated detection methods (horizontal axis) and all evaluated data streams (vertical axis). High error values are marked in red color; low values in blue. (Color figure online)

less frequent (3 and 5 in the entire stream), these errors' values will be higher due to the relatively greater distance for incorrect detections.

In the second measure, $D2$, a single nearest detection is searched for each drift. The $D2$ measure will penalize the methods that fail to recognize the actually occurring drifts or do so with a delay. The highest errors in this criterion are demonstrated by the DDM and MD3 methods, which sometimes signaled drifts with a delay or did not detect all changes.

The last measure, R, looks only at the actual number of drifts compared to the number of changes signaled by the method. In the case of too many detections, the measure will return values close to (but not exceeding) one. However, when the number of signaled changes is lower than the number of drifts, the measure will return values exceeding one. High error values are visible for the DDM and EDDM methods, but they do not exceed one, hence the methods were penalized for too many detections. However, the MD3 method obtained measure values exceeding one, which indicates too few detections on average, compared to the number of drifts. It is worth paying attention to the results for the CDDD method, which are exceptionally low in all streams for which the measure could be determined, but its problem is the tendency to not signal frequent and gradual drifts.

The results obtained from the second experiment were also subjected to a post-hoc *Nemenyi* test based on *Wilcoxon Signed Rank* across all three errors. The results for $D1$, $D2$, and R errors are presented in the Fig. 7. The overall statistical analysis did not consider the results of the CDDD detector due to the inability to calculate measures in the absence of detection by the method, which took place in 12 out of 24 calculations.

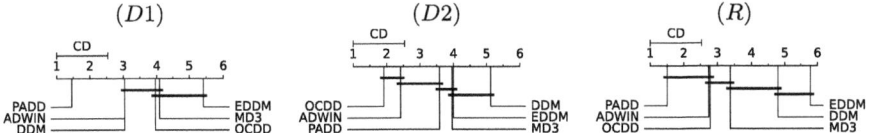

Fig. 7. Critical Difference diagrams for three drift detection error measures: D1 (left), D2 (center), and R (right). The CDDD method was not considered in the statistical evaluation due to the lack of possibility to calculate *drift detection measures* in certain data streams.

The results show that in the case of $D1$ error, the presented method is significantly the best lone leader of comparison. For the R error, the presented solution – while having the best ranking value – is statistically significantly co-dependent to the ADWIN and OCCD. For the $D2$, the best results were obtained by the OCDD method, and the PADD results are statistically dependent on the second method in terms of quality, the ADWIN method with the minimal difference in ranks.

It is worth emphasizing that those criteria should not be used independently to evaluate methods, and it is the juxtaposition of all three that describes the proper and effective method operation.

6 Conclusions and Future Works

This work proposed an unsupervised *Parallel Activations Drift Detector* that utilizes an untrained neural network to recognize significant changes in the posterior distribution of the data stream to signal concept drift according to the result of a statistical test stabilized by a pool of replications. The conducted experimental evaluation allowed to demonstrate that the proposed PADD method states a valuable tool in the context of reference methods. This enriches the pool of available drift detection methods, introducing an algorithm with high reliability in a proper configuration.

As part of future work, we plan to expand the hyperparameter calibration study – considering the introduction of a non-parametric version of the method interpreting the area under the curve of the threshold function of a statistical test – and a broader review of possible validators to replace the simple Student's T-test. An interesting area for developing the analysis would also be to research the dependencies between the outputs of the NN used, allowing for the naivety of their independence to be abandoned.

Acknowledgements. This work was supported by the statutory funds of the Department of Computer Systems and Networks, Wroclaw University of Science and Technology.

References

1. Agrahari, S., Singh, A.K.: Concept drift detection in data stream mining: a literature review. J. King Saud Univ. CISC **34**(10), 9523–9540 (2022)
2. Bach, S.H., Maloof, M.A.: Paired learners for concept drift. In: 2008 Eighth IEEE International Conference on Data Mining, pp. 23–32. IEEE (2008)
3. Baena-Garcıa, M., del Campo-Ávila, J., Fidalgo, R., Bifet, A., Gavalda, R., Morales-Bueno, R.: Early drift detection method. In: Fourth International Workshop on Knowledge Discovery from Data Streams, vol. 6, pp. 77–86. Citeseer (2006)
4. Bifet, A.: Classifier concept drift detection and the illusion of progress. In: Rutkowski, L., Korytkowski, M., Scherer, R., Tadeusiewicz, R., Zadeh, L.A., Zurada, J.M. (eds.) ICAISC 2017. LNCS (LNAI), vol. 10246, pp. 715–725. Springer, Cham (2017). https://doi.org/10.1007/978-3-319-59060-8_64
5. Bifet, A., Gavalda, R.: Learning from time-changing data with adaptive windowing. In: Proceedings of the 2007 SIAM International Conference on Data Mining, pp. 443–448. SIAM (2007)
6. Chen, T., Kornblith, S., Swersky, K., Norouzi, M., Hinton, G.E.: Big self-supervised models are strong semi-supervised learners. Adv. Neural. Inf. Process. Syst. **33**, 22243–22255 (2020)
7. Domingos, P., Hulten, G.: A general framework for mining massive data streams. J. Comput. Graph. Stat. **12**(4), 945–949 (2003)
8. Emmert-Streib, F., Yang, Z., Feng, H., Tripathi, S., Dehmer, M.: An introductory review of deep learning for prediction models with big data. Front. Artif. Intell. **3**, 4 (2020)
9. Feldmann, A., Gasser, O., Lichtblau, F., Pujol, E., Poese, I., Dietzel, C., Wagner, D., Wichtlhuber, M., Tapiador, J., Vallina-Rodriguez, N., et al.: A year in lockdown: how the waves of covid-19 impact internet traffic. Commun. ACM **64**(7), 101–108 (2021)
10. Gama, J., Medas, P., Castillo, G., Rodrigues, P.: Learning with Drift Detection. In: Bazzan, A.L.C., Labidi, S. (eds.) SBIA 2004. LNCS (LNAI), vol. 3171, pp. 286–295. Springer, Heidelberg (2004). https://doi.org/10.1007/978-3-540-28645-5_29
11. Gemaque, R.N., Costa, A.F.J., Giusti, R., Dos Santos, E.M.: An overview of unsupervised drift detection methods. Wiley Interdiscip. Rev. Data Min. Knowl. Discov. **10**(6), e1381 (2020)
12. Gözüaçık, Ö., Büyükçakır, A., Bonab, H., Can, F.: Unsupervised concept drift detection with a discriminative classifier. In: Proceedings of the 28th CIKM, pp. 2365–2368 (2019)
13. Gözüaçık, Ö., Can, F.: Concept learning using one-class classifiers for implicit drift detection in evolving data streams. Artif. Intell. Rev. **54**(5), 3725–3747 (2020). https://doi.org/10.1007/s10462-020-09939-x
14. Grzenda, M., Gomes, H.M., Bifet, A.: Delayed labelling evaluation for data streams. Data Min. Knowl. Disc. **34**(5), 1237–1266 (2020)
15. Hu, H., Kantardzic, M., Sethi, T.S.: No free lunch theorem for concept drift detection in streaming data classification: a review. Wiley Interdiscip. Rev. Data Min. Knowl. Discov. **10**(2), e1327 (2020)
16. Klikowski, J.: Concept drift detector based on centroid distance analysis. In: 2022 International Joint Conference on Neural Networks (IJCNN), pp. 1–8. IEEE (2022)
17. Klinkenberg, R., Renz, I.: Adaptive information filtering: learning in the presence of concept drifts. In: Learning for Text Categorization, pp. 33–40 (1998)

18. Komorniczak, J., Ksieniewicz, P.: Complexity-based drift detection for nonstationary data streams. Neurocomputing **552**, 126554 (2023)
19. Komorniczak, J., Zyblewski, P., Ksieniewicz, P.: Prior probability estimation in dynamically imbalanced data streams. In: 2021 International Joint Conference on Neural Networks (IJCNN), pp. 1–7. IEEE (2021)
20. Ksieniewicz, P., Zyblewski, P.: Stream-learn-open-source python library for difficult data stream batch analysis. Neurocomputing **478**, 11–21 (2022)
21. Li, B., Wang, Y.j., Yang, D.s., Li, Y.m., Ma, X.k.: FAAD: an unsupervised fast and accurate anomaly detection method for a multi-dimensional sequence over data stream. Front. Inf. Technol. Electron. Eng. **20**(3), 388–404 (2019)
22. Lindstrom, P., Mac Namee, B., Delany, S.J.: Drift detection using uncertainty distribution divergence. Evol. Syst. **4**, 13–25 (2013)
23. Liu, A., Lu, J., Liu, F., Zhang, G.: Accumulating regional density dissimilarity for concept drift detection in data streams. Pattern Recogn. **76**, 256–272 (2018)
24. Liu, S., et al.: Online active learning for drifting data streams. IEEE Trans. Neural Netw. Learn. Syst. **34**(1), 186–200 (2021)
25. Lu, J., Liu, A., Dong, F., Gu, F., Gama, J., Zhang, G.: Learning under concept drift: a review. IEEE Trans. Knowl. Data Eng. **31**(12), 2346–2363 (2018)
26. Minku, L.L., Yao, X.: DDD: a new ensemble approach for dealing with concept drift. IEEE Trans. Knowl. Data Eng. **24**(4), 619–633 (2011)
27. Montiel, J., Read, J., Bifet, A., Abdessalem, T.: Scikit-multiflow: a multi-output streaming framework. J. Mach. Learn. Res. **19**(72), 1–5 (2018)
28. Narkhede, M.V., Bartakke, P.P., Sutaone, M.S.: A review on weight initialization strategies for neural networks. Artif. Intell. Rev. **55**(1), 291–322 (2022)
29. Sethi, T.S., Kantardzic, M.: Don't pay for validation: detecting drifts from unlabeled data using margin density. Procedia Comput. Sci. **53**, 103–112 (2015)
30. Sethi, T.S., Kantardzic, M., Hu, H.: A grid density based framework for classifying streaming data in the presence of concept drift. J. Intell. Inf. Syst. **46**, 179–211 (2016)

Unsupervised Assessment of Landscape Shifts Based on Persistent Entropy and Topological Preservation

Sebastián Basterrech[(✉)]

Department of Applied Mathematics and Computer Science,
Technical University of Denmark, Kongens Lyngby, Denmark
sebbas@dtu.dk

Abstract. Concept drift typically refers to the analysis of changes in data distribution. A drift in the input data can have negative consequences on a learning predictor and the system's stability. The majority of concept drift methods emphasize the analysis of statistical changes in non-stationary data over time. In this context, we consider a slightly different perspective, where concept drift also integrates significant changes in the topological characteristics of the data stream. In this article, we introduce a novel framework for monitoring changes in multi-dimensional data streams. We explore variations in the topological structures of the data, presenting another angle on the standard concept drift. Our developed approach is based on persistent entropy and topology-preserving projections in a continual learning scenario. The framework operates in both unsupervised and supervised environments. To show the utility of the proposed framework, we analyze the model across three scenarios using data streams generated with MNIST samples. The obtained results reveal the potential of applying topological data analysis for shift detection and encourage further research in this area.

Keywords: Distribution shifts · Persistent entropy · Dimensionality reduction · Concept Drift · Self-organizing maps

1 Introduction

In continual learning scenarios, designing a machine learning (ML) model that is robust to distribution shifts is a crucial objective. Traditional ML methods are susceptible to data perturbations, and shifts in input data distribution can significantly affect the model's performance. Concept drift detectors encompass a family of techniques developed to analyze and detect distribution changes in the context of streaming data and time series. The concept is based on changes in the statistical characteristics of the data over time [18]. However, there are scenarios where certain modifications in the distributions are not relevant. For example, simple translations or scaling of the data may not provide meaningful information in some contexts. Therefore, it would be beneficial to define concept drift detectors that can detect distribution changes regardless of certain distortions or rotations of the data. There are objects that are essentially equivalent to each other if we consider "equivalence" in the sense that it is possible to define a simple continuous transformation that approximates one object to the other. The essence of an object remains unchanged under simple transformations, such as rotation, translation, scaling, and other types of continuous transformations [13]. On the other hand, there are objects that are essentially different, as it is not feasible to find any continuous transformation to transform one object into another [11]. Or at least, it is not easy to find such a transformation with low computational resources. The field of Topological Data Analysis (TDA) defines these concepts of equivalences and differences between objects through the mathematical formalism of algebraic topology. Persistent Entropy (PE) is a measure based in Shannon entropy that provides a summary of the geometric information derived from the topological features of a cloud of points [1,31]. It has been successfully used to effectively distinguish chaotic and periodic time series [31].

In this work, we introduce an extension of the classic concept drift that integrates algebraic topology. We extend the concept of drift, which is based on statistical and geometrical information, with another concept that incorporates the notion of *essential sameness* and *essential difference* [11]. Our work aims to provide insight into the research question: Is it adequate to identify a drift between two sequences if there exists a simple continuous bijective function that transforms one sequence into the other? In other words, is it adequate to identify a drift between two sequences when they are equivalent objects in terms of topology? We observe a drift when the geometric characteristics of a point cloud *essentially* change, becoming different from those of another point cloud. Figure 1 illustrates examples of objects that can be deformed using simple continuous transformations such as rotation, stretching, bending, and scaling. The first row of the figure displays three digits that are topologically equivalent. The second row also shows three topologically equivalent objects. It is not possible to transform any object from the first row into an object of the second row. To quantify these types of topological variations, we take into account concepts provided by TDA. We empirically investigate the changes in the distribution of high-dimensional data in an unsupervised scenario, using tools from persistence homology. We develop a general-purpose framework that projects the input data into a low-dimensional space using a dimensionality reduction technique that preserves the topological features of the data. We then apply metrics of persistent homology to evaluate significant changes. The projection from input space to latent space is made using Self-Organizing Maps (SOM), which is a mapping technique to reduce dimensionality while preserving the topological characteristics of the input space [25]. In the latent space, we explore the potential of persistent homology to find significant differences among data coming from consecutive chunks. We use the metric of persistent entropy, which summarizes the analysis of persistent homology in a single value [2]. In summary, this work offers the following contributions.

(i) We introduce a general-purpose framework for concept drift detection that operates in both supervised and unsupervised environments. This framework utilizes dimensionality reduction through a topological preserving mapping and evaluates significant topological changes using persistent homology.
(ii) The framework delivers results using a p-value score. When each chunk of data arrives, a non-parametric statistical test is performed, facilitating an easy monitoring of drifts. The hypothesis test is conducted on the values of persistent entropy. Since the framework provides a p-value score, the decision regarding the absence or presence of a drift is both robust and fast.
(iii) We provide an initial experimental evaluation with promising results across three case studies. We compare three dimensionality reduction techniques to evaluate the impact of preserving topological features when projecting data from the input space to the latent space. The results show the benefits of combining a topology-preserving mapping with information regarding persistent homology.

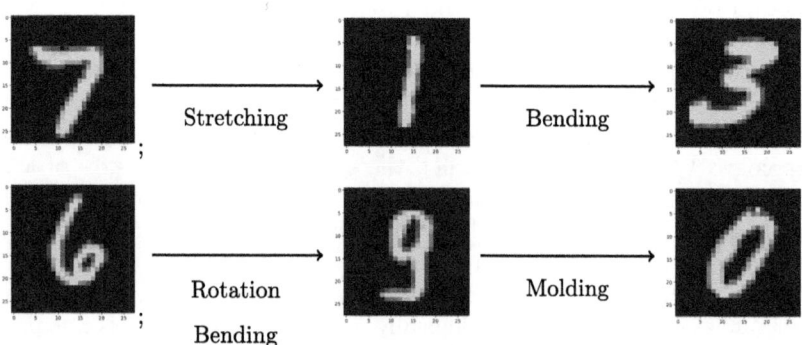

Fig. 1. Examples of objects that can be deformed and transformed into another object equivalent in terms of topology. However, some shapes cannot be considered equivalent because it is not possible to define a sequence of simple continuous transformations to deform an original shape into another one while maintaining the original structure. For instance, from any of the digits at the top of the figure, we cannot form any of the digits at the bottom of the figure.

2 Background

Drift detection techniques constitute a family of computational methods for detecting changes in the distribution of time series and data streams. Shifts in data distribution can occur in different forms. Probably, the most accepted categories are: sudden, gradual, and incremental shifts [42,45]. A majority of drift detection techniques employ a classifier to categorize incoming instances, and the predictor generates a class label for each input instance, which is then compared to the actual class label. Subsequently, the classifier's accuracy is used as a tool to determine whether a drift has occurred When the accuracy significantly decreases, this family of tools assumes that the data distribution has changed [17,19,23]. In this scenario, the effectiveness of various ensemble classifiers has been examined [14,26,27,30]. However, this approach can be applied only in a supervised context, and it requires the presence of ground truth labels, which are not always available. Another set of methods relies on empirically estimated distributions and statistical tests directly over the raw data [8,12,20,41]. These approaches may be sensitive to outliers and noise, and raw data analysis (e.g., applying density estimation) may also be affected by the curse of dimensionality [5,6]. Several studies propose to compare summary of statistics and aggregation metrics of the raw data, for instance, Cumulative Sum and Exponentially Weighted Moving Average [35]. For a more comprehensive review of the latest advances in the use of data descriptors for concept drift detection, see [15,22,33].

Persistent homology is a key instrument in TDA as it may be used to describe the inherent structure of complex objects such as manifolds [34]. Specifically, persistent homology studies the evolution of k-dimensional topological features (often referred to as *holes*) along a sequence of high-dimensional complex objects (named *simplicial complexes*) [1,11]. We understand topological features as shapes or data that remain unchanged under certain continuous transformations, such as connected components, independent cycles, and holes [11]. This process can be seen as tracking changes across filtrations at multiple scales, following a specific algorithm that analyzes the connectivity information among the data points [11]. Persistent Entropy, based in Shannon entropy, provides a summary of the information derived from persistent homology [1]. It is a measure for finding significant differences in the geometrical distribution of data points [1,32]. For a comprehensive and detailed exploration about TDA and persistent homology, see [11,32].

3 Methodology

This section outlines the contributions made in this brief article. First, we discuss the approach for transforming the input patterns into a different landscape that simplifies the analysis of drifts. Next, we introduce the process for estimating geometric changes between data points in different chunks. Finally, we present a global view of the developed framework as a pipeline of modules.

3.1 Creation of the Latent Space

Monitoring and detecting distribution shifts is specially harder in the case of high-dimensional data. Although some attempts have been introduced in the literature for sparse multivariate time series [39,46], the scalability of existing algorithms remains an issue. In particular, methods based on probability mass distribution encounter significant challenges in high-dimensional spaces. In addition, the computation of distances between vectors also has limitations in a high-dimensional space (e.g. Euclidean norm) [40], and a similarity analysis in the original input space may be computationally expensive. Consequently, it is often more resource-efficient to first convert the data into a latent space, and then carry out the similarity analysis. For this reason, a common approach is to transform the input data into a latent space, instead of making the analysis directly in the original space. We analyze an approach that projects the input points into a latent space using dimensionality reduction (DR) techniques, which is a common method for handling data in high dimensions. Here, we investigate the projections generated by Self Organizing Maps (SOMs) (also called Kohonen networks), and we compare the results with other two popular DR models (a linear projection (PCA) and the Kernel PCA). The selection of an adequate data descriptor is crucial for ensuring a proper geometry in the latent space preserving the main features of the original space.

SOM is a bio-inspired method that combines concepts from Hebbian learning, vector quantization, and competitive learning [3,25]. Real-world data most often contain redundancies and inherent correlations among the variables. SOM is a two-layered neural network that transforms intricate relationships among high-dimensional data into straightforward geometric relationships on a standard lattice, typically a two-dimensional grid [16]. Despite its simplicity, the SOM model is effective as a DR method, a clustering method, and a visualization tool for high-dimensional data [36]. Another advantage is that the method is applicable to unsupervised problems and has the capability to preserve the most important topological features of the reference data [25].

3.2 Assessing Shifts in the Latent Space

Recently, it was introduced a clustering method based in SOM for assessing distribution changes in data streams with high dimensional data [3]. SOM is used for projecting the input data into a latent space, then the analysis is done in the latent space, where the authors computed a distance matrix between the input pattern and cluster centers. The assessment of the distribution shifts is done by applying a statistical summary. This approach of using a data descriptor was also applied in [20,21], and is commonly used in methods based on kernel projections [38]. Here, we modify the framework introduced in [3], which is based on distances and statistical summaries of the points in the latent space, to an approach that assesses topological changes according to the homological characteristics of the points in the latent space. Once the DR mapping is done, a distance matrix per each projected point is computed. The distance matrix has the information between the projected point and the cluster centers. We use relative locations instead of directly working with the coordinates of points in the latent space generated by the DR method. The coordinates are arbitrarily selected and often do not consider any property of the data itself. There are even problems in cases where the coordinates are not natural in any sense [10]. Therefore, the relative locations of the point cloud in the latent space are computed by calculating the distances between the mass centers and the projected points. Hence, our focus is on the geometric properties of the latent space, independent of the chosen coordinates in the latent space. This methodological approach is illustrated in Fig. 2. Note that, the approach is general in the sense that any type of DR technique can be used.

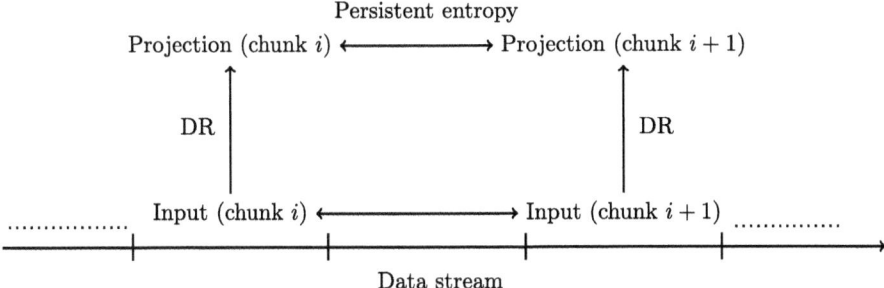

Fig. 2. Assessing the topological changes in the latent space: comparing the persistent entropy of projected points using a Dimensionality Reduction (DR) technique.

3.3 Pipeline of the Proposed Framework

The proposed framework for assessing topological changes based in persistent entropy involves the following steps:

i) **Initial setup**: An initial batch of points is used as a pre-phase procedure to train the selected DR method.

ii) **Dimensionality reduction**: When a chunk of data arrives, then the input data points are projected using the chosen DR method.

iii) **Embedding of the geometrical properties in the latent space**: For each projected data point, the distance between the projected point and each centroid is computed. By centroid, we refer to the mass center of a cluster, which can be thought of as the average position of all the points within the cluster, or a representative data point that best exemplifies the cluster's characteristics. Then, for each chunk, a matrix is created (with dimensions of number of clusters × Chunk size) where the column vector has the distances between a projected point and the centroids.

iv) **Representation of topological features**: A persistent diagram is created using the distance matrix described in the previous step. The persistence diagram summarizes the information in the distance matrix (note that per each projected point is computed a distance matrix). Then, the persistent entropy is computed from the collection of persistence diagrams within the chunk, ignoring the infinity bar [37]. The representation of topological features returns a real vector with a number of elements equal to the chunk size.

v) **Statistical analysis**. We compute a final index that corresponds to a comparison between the representation of topological features in the current chunk and the representation of topological features in a reference chunk. In our experiments, we compare two consecutive chunks. The comparison between these two vectors is done using a non-parametric test. In our experiments, we applied the Mann-Whitney U test. However, analysis of similarity between these two vectors can be done using other tools. The statistical test provides a p-value score, which we use for monitoring changes between a current chunk of data and a reference chunk of data.

Finally, when the global procedure is applied online, it generates a sequence of p-value scores. This sequence provides information about significant changes in the topological properties of the data stream. Note, an initial training phase is performed to compute the initial clustering and its representative mass centers. An initial time window of the data stream is used for training the SOM weights and other global parameters. After this initial training phase, we continue the learning process following a usual continual learning scenario. The procedure described previously can include an additional step that involves adjusting the DR technique once a shift is detected. When a shift is detected, the SOM method can be re-trained using an arbitrary batch of data (for example, the last chunk or the last n chunks of data), with the precautions to avoid catastrophic forgetting. This online calibration procedure was not investigated in this initial empirical exploration and could be a promising direction for future research.

4 Experimental Results

In this section, we explore the utility of the proposed approach for monitoring and detecting topological changes in a data stream in the context of continual learning. We designed the experiments to evaluate and contrast the efficacy of the dimensionality reduction methods previously discussed: PCA, Kernel PCA, and SOM. There exists a deficiency in the availability of extensive and diverse real-world data streams within high-dimensional spaces for analyzing the impact of distribution changes [42]. This inconvenience is more relevant in the domain of unsupervised analysis of streaming data. As a consequence, we created three synthetic data streams with annotated shifts. Our analysis is unsupervised (we do not use the data labels); we use the information about the annotated shifts to evaluate the performance of the proposed approach. The created stream has samples from MNIST [29]. The annotations indicate the time stamps where the topological differences between two point clouds were injected.

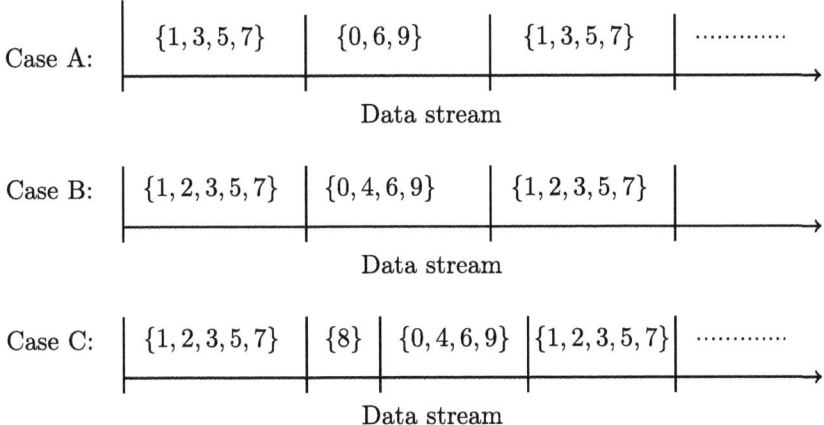

Fig. 3. Creation of synthetic case studies. Data streams were generated with the MNIST samples interchanged among the different topological types. The graphics illustrate the transition between the sequence of images from one topological type to another type.

4.1 Benchmark Data

We generated three synthetic datasets using the MNIST dataset [28], following the methodology outlined in [3,4,33]. The procedure consists of creating a stream with chunks of samples that follow a specific distribution, and then alternating these chunks with chunks of instances from a different distribution. By construction, the time stamps of the distribution changes are arbitrary predefined. As a consequence, we have marked the exact time stamps at which a drift was injected to evaluate the capacity of the approach in monitoring shifts. We created the data to check if the method can detect "essential" changes between sequences of digits. We divided the digits into three groups: those without any holes (zero-dimensional homology), those with one hole (one-dimensional homology), and those with two holes. We denote the three experimental studies as: A, B and C. For the three cases, we analyze 20000 samples. For case studies A and B, the drift is injected every 1000 samples, and for scenario C, the drift occurs every 500 samples. In case study A, we analyze a data stream where the changes occur between chunks with digits in $\{1,3,5,7\}$ (zero-dimensional homology) and chunks with digits in $\{0,6,9\}$ (one-dimensional homology). Case study B also includes the digits 2 and 4. These digits are problematic due to variations in handwriting. Some individuals write the numbers 2 and 4 without any hole, while others write them with one hole, depending on individual handwriting style. Then, case study B has a data stream considering the following exchanges between points in $\{1,2,3,5,7\}$ and $\{0,4,6,9\}$. Finally, in case study C, we evaluate a data stream that includes the number 8, which is not topologically equivalent to any of the other digits. Case C exchanges samples from the three subsets $\{1,2,3,5,7\}$, $\{0,4,6,9\}$, and $\{8\}$. Figure 3 illustrates the process by which the streaming data was generated for each of the three scenarios. Case study A and B have 20 exchanges between subsets containing digits with and without holes, and case study C involves 40 exchanges between subsets containing digits without holes, with one hole, and two holes.

4.2 Experimental Settings

For each of the three case studies, we used the first 20% of samples for the initial setup, and we train the parameters of the SOM algorithm. This training was made offline, as a pre-phase of the continual learning process. The SOM algorithm has a grid with 10×10 neurons. We also analyzed three values for the chunk size parameter $\{50, 100, 250\}$. The quality assessment of the monitoring for the shifts was done using a p-value computed with the non-parametric Mann-Whitney U test. We also evaluate the approach by applying PCA and Kernel-PCA instead of SOM, i.e. we apply PCA and Kernel-PCA in step (ii) of the procedure described in Sect. 3.3. The same initial time window used for SOM was applied to configure PCA and Kernel PCA. Then, we apply the trained DR methods to project the data points into a latent space, following a CL setting.

4.3 Implementation Details

The implementation of the methods developed during our investigation, as well as the experimental environment, was carried out using the *Python* v3.9 programming language. Several libraries were utilized to facilitate this process, including *NumPy* v1.19.5 for numerical computations, *stream-learn* v0.8.16 for handling data streams, *Sklearn* v1.0.2 for machine learning tasks [9], and the *Persim* v0.3.6 package for operations related to persistent homology [7,37,44]. The source code of our investigation and the datasets are available in the git repository[1].

4.4 Results

Figure 4 illustrates the latent space generated by the SOM projections of the data in case C. The curve (blue dots) in the figure represents the mean distance between the projected points and the mass centers of the clusters. It shows the evolution of the mean values for each of these distance matrices. In addition, Fig. 4 shows the results of applying Pruned Exact Linear Time (PELT) algorithm [24] to detect changes in the sequence of these mean values. The PELT method is recognized for its computational efficiency, when compared with other change-point detection techniques [43]. This experiment was done offline with the purpose of visualizing the complexity of the problem. The aim was to show that detecting shifts in the generated data is not sufficient by merely applying a DR method and computing the mean distance to the mass centers. We consider the blue curve in Fig. 4 as an illustration of the data stream characteristics for

---

[1] https://github.com/sebabaster/Drift-persistence.

case study C. This figure also shows the detected change points using the PELT technique. The background colors represent the changes detected by PELT, and the vertical green lines indicate the injected shifts. The visualization using PELT serves as a representation of the problem's complexity. It was conducted independently of the other experiments, in which we compare the approaches using SOM, PCA, and Kernel-PCA. In other words, we are not comparing PELT to the other techniques; this experiment was conducted to evaluate whether the generated data presents difficulties for a well-known technique for change point detection. The other experiments simulate a continual learning environment, making them even more complex than when the problem is addressed offline. Figure 6 presents the results for the three DR techniques. These techniques were evaluated using case study C. The data stream was split into chunks of 250 samples. The vertical lines (represented by green dashed lines) indicate the injected shifts. The horizontal line represents a p-value of 0.05. According to the results, the linear projection is unable to accurately predict the shifts. This observation is consistent with other studies in the literature that discuss the limitations of linear projections in detecting distribution shifts [3,22]. The performance of SOM appears to be slightly better to that of Kernel-PCA. For instance, refer to the p-values in the chunks between 30 and 35. Chunk size is a crucial parameter for experiments conducted in online settings. The influence of the chunk size is shown in Figs. 7, 8, and 9, as well as in Table 1. Figure 7 shows the results of using PCA in the step (ii) described in Sect. 3.3. There are two graphs: the top graph shows the results for chunks with 50 instances, and the bottom graph shows the results for chunks of 100 samples. This figure also highlights the limitations of linear projections for solving this specific problem, as the method provides few alarms and detects only a low number of drifts. Figure 8 presents the results of Kernel PCA for chunks with 50 and 100 samples. Similar graphics are depicted in Fig. 9 where the results of SOM for chunks with 50 and 100 samples are presented. Table 1 summarizes the results of our experiments, including an evaluation of the impact of chunk size. The last two columns show the flags generated by the model using p-values with significance levels of 0.05 and 0.1. Additionally, we show the number of injected drifts in the experiment. Note that this number is an approximation due to the anomalies that can exist in the datasets (digits with a different number of holes than expected). According to the table, it seems that SOM may provide results using p-values with a 0.05 level of significance, while Kernel-PCA obtains better results when a p-value with a 0.1 level of significance is considered. As one might intuitively expect, smaller chunks decrease the quality of geometric pattern analysis. In contrast, larger chunks can encompass more than one shift. The optimal chunk size should be determined through experimentation, taking this trade-off into account.

5 Discussion

As far as we know, this is the first work that attempts to define drifts, including sudden changes in the topological features of the data. Our research hypothesis was to investigate whether the use of TDA may be helpful or not for monitoring topological changes and detecting "essential" differences between chunks of objects. In other words, if we have a data stream with different types of doughnuts and then suddenly the data contains coffee cups, a "traditional" drift detector would detect a drift (using the standard notion of drift [18]). However, if we consider purely topological information, both shapes (coffee cups and doughnuts) are considered equivalent. Therefore, it would not be appropriate to consider this type of changes as a drift; in the case that we are interested in detecting "essential" differences. Shapes with the same Euler characteristic cannot be considered as "essentially" different. The inclusion of TDA in the concept drift domain may be highly beneficial for avoiding false alarm when the data has only suffer simple transformations such as scaling, distortions and rotations.

To answer this research question, we implemented a set of experiments based in the MNIST dataset [28]. We defined three case studies based on changes in the Euler characteristic of the data. The digits were not individually visually inspected; instead, they were sampled using a uniform distribution. Consequently, there may be instances where an image was expected to have a specific number of holes, but the sample shows a different number. We illustrate these anomaly examples in Fig. 5. The figure contains four graphics. The first image has at least two holes, whereas the expected number is one. The second image has one hole, but it represents either the number three or five, both of which typically have zero holes. The third image represents the number six, but it was not properly drawn, so the circle is not closed. Finally, we also show an image with pixels that are not connected, forming more than one connected component. Therefore, the analyzed scenarios contain noisy marks in the stream. For further analyses, it would be beneficial to generate new datasets with higher quality annotated data streams. In this study, we did not evaluate different strategies to mitigate the catastrophic forgetting problem. When a drift is detected, it is necessary to define an appropriate strategy to fine-tune and update the framework. This is especially important for SOM. We left this evaluation for future research. Another limitation of our experimental analysis is that the framework is composed of two modules: dimensionality reduction and persistent entropy. Further investigation is required to analyze the relevance of each of these modules and how they affect drift detection. Additionally, an ablation study considering the different steps in the pipeline would enhance the reliability of our experiments.

Last but not least, defining concept drift solely based on changes in topological features appears to present an incomplete understanding of the phenomenon. Analyzing objects only based on their topological aspects ignores the meaning of the objects. Human interpretation, including contextual insights and semantic understanding of the information, may also be relevant in some problems. For example, 6 and 9 are different digits according to semantic understanding, but a simple rotation can transform one digit into the other. Therefore, to define drift ignoring human interpretation, which is the case of purely consider topology also have limitations. Combining the three aspects–statistics and geometry (both considered in previous literature), and topology (the approach introduced here)–in

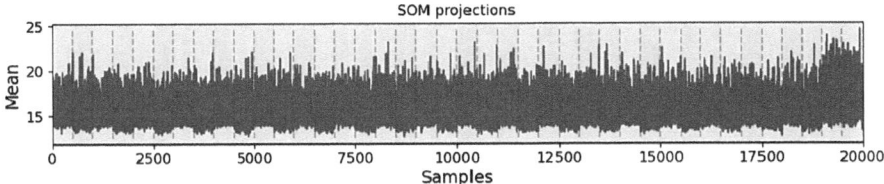

Fig. 4. Example of the latent space. Off-line analysis of the latent space generated by the SOM projections, and applying change point detection over the distance matrix. (Color figure online)

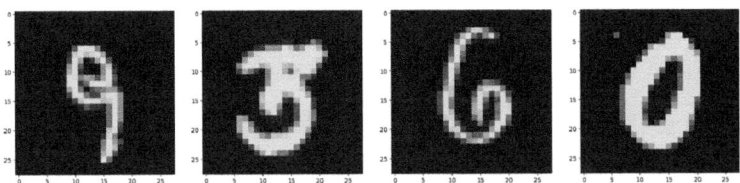

Fig. 5. An example of digits with a structure different from the one assumed in the case studies. The first image has two holes instead of one. The second image, which can be a number 3 or 5, has a hole. The third image doesn't have any holes. The last image is not a connected component (it has an isolated pixel in the top-left).

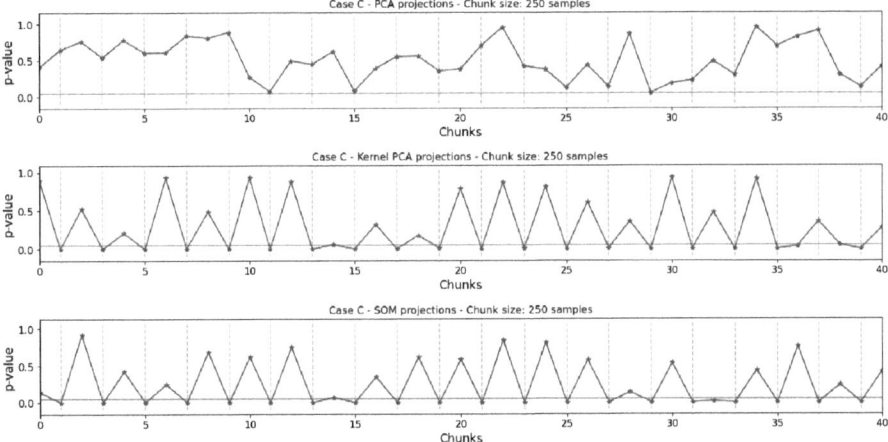

Fig. 6. Comparison among the three projections. The comparison was made over the dataset C with chunk size of 250 samples. (Color figure online)

the definition of drifts could be a promising direction. An additional interesting angle would be to also integrate the meaning of the information. Therefore, there are several avenues for future exploration, presenting many open questions that remain to be addressed.

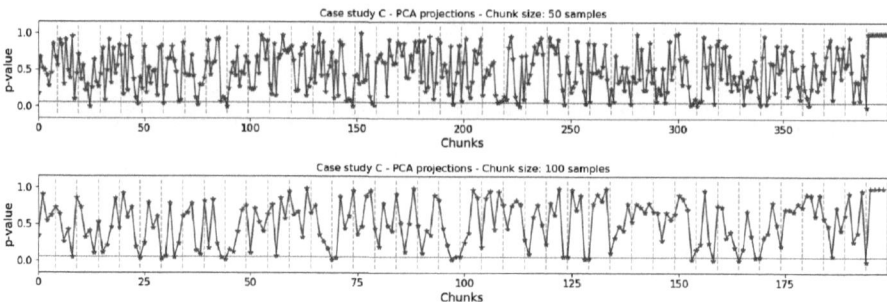

Fig. 7. Results of the method using PCA projections in case study C for chunk sizes of 50 and 100 samples.

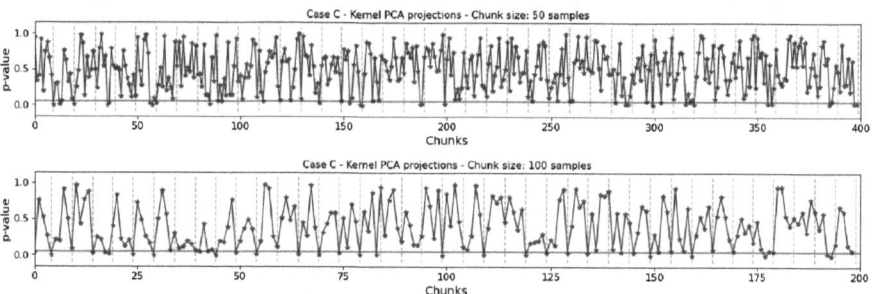

Fig. 8. Results of the method using Kernel PCA projections in case study C for chunk sizes of 50 and 100 samples.

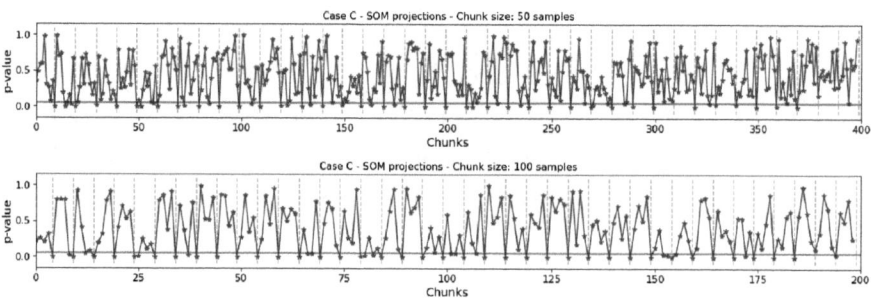

Fig. 9. Results of the method using SOM projections in case study C for chunk sizes of 50 and 100 samples.

Table 1. Sensitive analysis of the chunk size for the three study cases, the three methods, and two levels of significance of the hypothesis test (0.05 and 0.1). The table shows the number of potential drifts (flags) detected by each method. The fourth column shows the number of injected drifts. This is an approximate number because, during the construction of the dataset, each image wasn't individually verified. Consequently, digits like 2, 4, 6, and 9 may sometimes have a hole and sometimes not.

Case study	Chunk size	Method	Drifts	Flags (p-value at 0.05 level)	Flags (p-value at 0.1 level)
A	50	SOM	20	13	30
		PCA		22	51
		KernelPCA		27	58
	100	SOM	20	9	16
		PCA		10	27
		KernelPCA		14	22
	250	SOM	20	3	5
		PCA		6	11
		KernelPCA		4	8
B	50	SOM	20	18	36
		PCA		17	34
		KernelPCA		17	39
	100	SOM	20	10	26
		PCA		6	12
		KernelPCA		10	21
	250	SOM	20	4	12
		PCA		3	4
		KernelPCA		9	14
C	50	SOM	40	53	75
		PCA		21	44
		KernelPCA		27	51
	100	SOM	40	43	53
		PCA		13	26
		KernelPCA		26	37
	250	SOM	40	40	43
		PCA		8	10
		KernelPCA		25	35

6 Conclusions and Future Work

We introduced a novel approach to concept drift detection, leveraging algebraic topology and persistent entropy. We broaden the scope of concept drift, which is typically associated solely with statistical distribution changes in incoming data. In the explored approach, we also integrate the definition of drift to detect significant changes in topological features. The framework uses the SOM algorithm to transform the input space, reducing its dimensionality while preserving its topological characteristics. Then, the analysis of data drifts is conducted in the latent space. We explore the potential of persistent entropy to identify significant differences among data from consecutive chunks. We showed the performance of the method over three case studies (based on the MNIST dataset), and we compared the performance with PCA and Kernel-PCA. The proposed method does not make any assumption about the data distribution. Additionally, it can be applied to both supervised and unsupervised problems. We believe that this work is an initial step towards applying TDA in the area of concept drift. A potential direction for future research could involve evaluating the framework with different types of data streams, including dynamic graphs. Furthermore, it would be interesting to compare persistent entropy with other measures.

References

1. Atienza, N., Gonzalez-Diaz, R., Rucco, M.: Persistent entropy for separating topological features from noise in vietoris-rips complexes. J. Intell. Inf. Syst. **52**(3), 637–655 (2017). https://doi.org/10.1007/s10844-017-0473-4
2. Atienza, N., Gonzalez-Díaz, R., Soriano-Trigueros, M.: On the stability of persistent entropy and new summary functions for topological data analysis. Pattern Recogn. **107**, 107509 (2020). https://doi.org/10.1016/j.patcog.2020.107509
3. Basterrech, S., Clemmensen, L., Rubino, G.: A self-organizing clustering system for unsupervised distribution shift detection. In: 2024 International Joint Conference on Neural Networks (IJCNN), pp. 1–9 (2024). https://doi.org/10.1109/IJCNN60899.2024.10650314
4. Basterrech, S., Kasprzak, A., Platos, J., Wozniak, M.: A Continual Learning System with Self Domain Shift Adaptation for Fake News Detection. In: 10th IEEE International Conference on Data Science and Advanced Analytics, DSAA 2023, Thessaloniki, Greece, 9-13 October 2023, pp. 1–10. IEEE (2023). https://doi.org/10.1109/DSAA60987.2023.10302539
5. Basterrech, S., Platoš, J., Rubino, G., Woźniak, M.: Experimental Analysis on Dissimilarity Metrics and Sudden Concept Drift Detection. In: Abraham, A., Pllana, S., Casalino, G., Ma, K., Bajaj, A. (eds.) Intelligent Systems Design and Applications, pp. 190–199. Springer Nature Switzerland, Cham (2023). https://doi.org/10.1007/978-3-031-35501-1_19
6. Basterrech, S., Woźniak, M.: Tracking changes using Kullback-Leibler divergence for the continual learning. In: 2022 IEEE International Conference on Systems, Man, and Cybernetics (SMC), pp. 3279–3285 (2022). https://doi.org/10.1109/SMC53654.2022.9945547
7. Bauer, U.: Ripser: efficient computation of Vietoris–Rips persistence barcodes. J. Appl. Comput. Topol. **5**(3), 391–423 (2021). https://doi.org/10.1007/s41468-021-00071-5
8. Blanco, I.I.F., del Campo-Avila, J., Ramos-Jimenez, G., Bueno, R.M., Diaz, A.A.O., Mota, Y.C.: Online and non-parametric drift detection methods based on hoeffding's bounds. IEEE Trans. Knowl. Data Eng. **27**(3), 810–823 (2015)

9. Buitinck, L., et al.: API design for machine learning software: experiences from the scikit-learn project. In: ECML PKDD Workshop: Languages for Data Mining and Machine Learning, pp. 108–122 (2013)
10. Carlsson, E., Carlsson, G., Silva, V.: An algebraic topological method for feature identification. Int. J. Comput. Geometry Appl. **16**, 291–314 (2006). https://doi.org/10.1142/S021819590600204X
11. Carlsson, G.: Topology and data. Bull. Am. Math. Soc. **2**, 255–308 (2009)
12. Clemmensen, L.H., Kjærsgaard, R.D.: Data representativity for machine learning and AI systems. arXiv preprint arXiv:2203.04706 (2023)
13. Dai, M., Kurtek, S., Klassen, E., Srivastava, A.: Statistical shape analysis, pp. 1197–1211. Springer International Publishing, Cham (2021). https://doi.org/10.1007/978-3-030-63416-2_778
14. Du, L., Song, Q., Zhu, L., Zhu, X.: A selective detector ensemble for concept drift detection. Comput. J., 457–471 (2015). https://doi.org/10.1093/comjnl/bxu050
15. Faber, K., Corizzo, R., Sniezynski, B., Baron, M., Japkowicz, N.: WATCH: wasserstein change point detection for high-dimensional time series data. In: 2021 IEEE International Conference on Big Data (Big Data), pp. 4450–4459 (2021). https://doi.org/10.1109/BigData52589.2021.9671962
16. Ferles, C., Papanikolaou, Y., Naidoo, K.J.: Denoising autoencoder self-organizing map (DASOM). Neural Netw. **105**, 112–131 (2018). https://doi.org/10.1016/j.neunet.2018.04.016
17. Gama, J., Medas, P., Castillo, G., Rodrigues, P.: Learning with drift detection. In: Bazzan, A.L.C., Labidi, S. (eds.) SBIA 2004. LNCS (LNAI), vol. 3171, pp. 286–295. Springer, Heidelberg (2004). https://doi.org/10.1007/978-3-540-28645-5_29
18. Gama, J., Žliobaitė, I., Bifet, A., Pechenizkiy, M., Bouchachia, A.: A survey on concept drift adaptation. ACM Comput. Surv. (CSUR) **46**(4), 44 (2014)
19. Gonzalvez, P.M., de Carvalho Santos, S., Barros, R., Vieira, D.: A comparative study on concept drift detectors. Expert Syst. Appl. **41**(18), 8144–8156 (2014)
20. Hinder, F., Artelt, A., Hammer, B.: Towards non-parametric drift detection via dynamic adapting window independence drift detection (DAWIDD). In: Proceedings of the 37th International Conference on Machine Learning. ICML 2020 (2020). https://jmlr.org/
21. Hinder, F., Vaquet, V., Brinkrolf, J., Hammer, B.: Model-based explanations of concept drift. Neurocomputing **555**, 126640 (2023). https://doi.org/10.1016/j.neucom.2023.126640
22. Hinder, F., Vaquet, V., Hammer, B.: Suitability of different metric choices for concept drift detection. In: Bouadi, T., Fromont, E., Hüllermeier, E. (eds.) Advances in Intelligent Data Analysis XX, pp. 157–170. Springer International Publishing, Cham (2022). https://doi.org/10.1007/978-3-031-01333-1_13
23. Goldenberg, I., Webb, G.I.: Survey of distance measures for quantifying concept drift and shift in numeric data. Knowl. Inf. Syst. **60**(2), 591–615 (2018). https://doi.org/10.1007/s10115-018-1257-z
24. Killick, R., Fearnhead, P., Eckley, I.A.: Optimal detection of changepoints with a linear computational cost. J. Am. Stat. Assoc. **107**(500), 1590–1598 (2012). https://doi.org/10.1080/01621459.2012.737745
25. Kohonen, T.: Essentials of the self-organizing map. Neural Netw. **37**, 52–65 (2013). https://doi.org/10.1016/j.neunet.2012.09.018
26. Kolter, J., Maloof, M.: Dynamic weighted majority: a new ensemble method for tracking concept drift. In: Data Mining, 2003. ICDM 2003. Third IEEE International Conference on, pp. 123 – 130 (2003)

27. Lapinski, A., Krawczyk, B., Ksieniewicz, P., Wozniak, M.: An empirical insight into concept drift detectors ensemble strategies, pp. 1–8 (2018). https://doi.org/10.1109/CEC.2018.8477962
28. Le Cun, Y., Bottou, L., Bengio, Y., Haffner, P.: Gradient-based learning applied to document recognition. Proc. IEEE **86**(11), 2278–2324 (1998)
29. LeCun, Y., Cortes, C.: MNIST handwritten digit database (2010). http://yann.lecun.com/exdb/mnist/
30. Maciel, B.I.F., Santos, S.G.T.C., Barros, R.S.M.: A lightweight concept drift detection ensemble. In: 2015 IEEE 27th International Conference on Tools with Artificial Intelligence (ICTAI), pp. 1061–1068 (Nov 2015)
31. Myers, A., Munch, E., Khasawneh, F.A.: Persistent homology of complex networks for dynamic state detection. Phys. Rev. E **100**, 022314 (2019). https://doi.org/10.1103/PhysRevE.100.022314
32. Oner, D., Garin, A., Koziński, M., Hess, K., Fua, P.: Persistent homology with improved locality information for more effective delineation. IEEE Trans. Pattern Anal. Mach. Intell. **45**(8), 10588–10595 (2023). https://doi.org/10.1109/TPAMI.2023.3246921
33. Rabanser, S., Gunnemann, S., Lipton, Z.C.: Failing loudly: an empirical study of methods for detecting dataset shift. In: 33rd Conference on Neural Information Processing Systems (NeurIPS 2019), Vancouver, Canada (2019)
34. Rieck, B., et al.: Neural Persistence: a complexity measure for deep neural networks using algebraic topology. In: International Conference on Learning Representations (2019). https://openreview.net/forum?id=ByxkijC5FQ
35. Ross, G.J., Adams, N.M., Tasoulis, D.K., Hand, D.J.: Exponentially weighted moving average charts for detecting concept drift. Pattern Recogn. Lett. **33**(2), 191–198 (2012)
36. Saraswati, A., Nguyen, V.T., Hagenbuchner, M., Tsoi, A.C.: High-resolution self-organizing maps for advanced visualization and dimension reduction. Neural Netw. **105**, 166–184 (2018). https://doi.org/10.1016/j.neunet.2018.04.011
37. Saul, N., Tralie, C.: Persim: A Python package for persistence diagrams (2019). https://persim.scikit-tda.org/
38. Schölkopf, B., Smola, A.: Learning with Kernels: Support Vector Machines, Regularization, Optimization, and Beyond. MIT Press, Adaptive computation and machine learning (2002)
39. Shang, M., Yuan, Y., Luo, X., Zhou, M.: An α-β-divergence-generalized recommender for highly accurate predictions of missing user preferences. IEEE Trans. Cybern. (2022). https://doi.org/10.1109/TCYB.2020.3026425
40. Sobczyk, A., Luisier, M.: Approximate Euclidean lengths and distances beyond Johnson-Lindenstrauss. In: Koyejo, S., Mohamed, S., Agarwal, A., Belgrave, D., Cho, K., Oh, A. (eds.) Advances in Neural Information Processing Systems, vol. 35, pp. 19357–19369. Curran Associates, Inc. (2022)
41. Sobolewski, P., Wozniak, M.: Concept drift detection and model selection with simulated recurrence and ensembles of statistical detectors. J. Univ. Comput. Sci. **19**(4), 462–483 (2013)
42. Souza, V.M.A., dos Reis, D.M., Maletzke, A.G., Batista, G.E.A.P.A.: Challenges in benchmarking stream learning algorithms with real-world data. Data Min. Knowl. Disc. **34**(6), 1805–1858 (2020). https://doi.org/10.1007/s10618-020-00698-5
43. Svoboda, R., Basterrech, S., Kozal, J., Platoš, J., Woźniak, M.: A natural gas consumption forecasting system for continual learning scenarios based on hoeffding trees with change point detection mechanism. Knowl. Based Syst. **304**, 112482 (2024). https://doi.org/10.1016/j.knosys.2024.112482

44. Tralie, C., Saul, N., Bar-On, R.: Ripser.py: A lean persistent homology library for Python. J. Open Source Softw. **3**(29), 925 (2018). https://doi.org/10.21105/joss.00925
45. Wozniak, M., Zyblewski, P., Ksieniewicz, P.: Active weighted aging ensemble for drifted data stream classification. Inf. Sci. **630**, 286–304 (2023). https://doi.org/10.1016/j.ins.2023.02.046
46. Zhang, J., Wei, Z., Yan, Z., Zhou, M., Pani, A.: Online change-point detection in sparse time series with application to online advertising. IEEE Trans. Syst. Man Cybern. Syst. **49**(6), 1141–1151 (2019). https://doi.org/10.1109/TSMC.2017.2738151

Addressing Temporal Dependence, Concept Drifts, and Forgetting in Data Streams

Federico Giannini(✉) and Emanuele Della Valle

DEIB, Politecnico di Milano, Milano, Italy
{federico.giannini,emanuele.dellavalle}@polimi.it

Abstract. Several challenges arise when applying Machine Learning in data streams. Firstly, data points are continuously generated, and algorithms should continuously learn from each. The solution should also promptly adapt to changes in data distribution (known as concept drifts). Additionally, while addressing concept drifts, preserving the knowledge gained from past data is crucial to avoid the problem of catastrophic forgetting. Finally, one should consider the temporal dependence that data points may exhibit. Three communities address these problems separately: Streaming Machine Learning (SML), Continual Learning (CL), and Time Series Analysis (TSA). In our previous research, we proposed Continuous Progressive Neural Networks (cPNN), a first approach that considers all the challenges together. It bridges SML, CL, and TSA by producing a continuous adaptation of the CL strategy of Progressive Neural Networks. It uses transfer learning to adapt to changes quickly and manages temporal dependence using Long Short-Term Memory. In this work, we present a comprehensive experimental campaign that analyzes the behaviour of SML models and cPNN in the case of complex temporal dependence and various concept drifts on synthetic and real data streams. Results bring statistical evidence that SML models struggle with substantial temporal dependence, while cPNN is a viable solution.

Keywords: Data streams · Concept drifts · Temporal dependence

1 Introduction

Applying Machine Learning models to classify data points brings about a unique set of challenges in a scenario where data is generated from an unbounded data stream [3]. Firstly, a one-time learning phase on the entire dataset is impossible as the data is not available at once. New data points arrive continuously, and the model must learn from them. We want, therefore, to continuously train the model on single mini-batches or even a single data point at a time. Secondly, the distribution of data may change over time. This phenomenon is known as *concept drift* [17]. The goal is quickly adapting to changes to avoid performance slumps after a concept drift. Furthermore, data points may be autocorrelated, with each data point potentially relying on the preceding ones at a given timestamp [25]. Considering this dependence, it is crucial to solve the classification tasks efficiently. Finally, when a concept drift occurs, the model must learn the new distribution, which can lead to losing its predictive ability on previous ones. The stability-plasticity dilemma [19] regulates this phenomenon where plasticity is the ability to learn new knowledge, while stability is remembering the past acquired one. Too much plasticity could lead to *catastrophic forgetting* [14]. Conversely, too much stability can lead to difficulty in learning new knowledge. This problem may be particularly relevant when old distributions reoccur over time, or past knowledge may be useful to solve the problem associated with the current distribution.

These challenges are addressed by three main communities. Streaming Machine Learning (SML) [3] aims to enable continuous learning and adapt quickly to concept drifts. However, it assumes that data is independent and identically distributed (shortly i.i.d.) between two concept drifts. It, thus, fails to account for temporal dependence. Additionally, forgetting is usually ignored. Although Time Series Analysis [5], and particularly Recurrent Neural Networks (RNN), offer a potential solution for the former, they are poorly used in the streaming context. Moreover, neural networks may catastrophically forget the knowledge of previous concepts. Long Short-Term Memory (LSTM) [13] is one of this domain's most commonly used recurrent RNN architectures. Finally, Continual Learning (CL) [16] focuses on avoiding forgetting in Deep Learning models but assumes the data stream to be composed of large batches of data points accessible simultaneously. In this context, Progressive Neural Networks (PNN) [23] is a strategy to deal with forgetting and uses transfer learning to recycle old knowledge. Given these premises, Continuous Progressive Neural Networks (cPNN) [9] acts as a bridge between SML, CL and TSA and represents the first pioneering solution to solve all these challenges simultaneously. It manages temporal dependence using cLSTM, a continuous version of LSTM that buffers the data stream in fixed-size mini-batches and builds sequences on each. It then applies a PNN on top of cLSTM, allowing the solution to avoid forgetting and exploit transfer learning to adapt to new concepts quickly. cPNN is a *periodic classifier* that requires mini-batches of data points to perform inference and training.

This work presents a complete analysis of SML models and cPNN to answer the following **research questions**: *How do SML models and cPNN react to different types of drifts? Are they able to manage temporal dependence?* We set up an experimental campaign involving synthetic and real data streams. Additionally, the classification problems implicate elaborated temporal dependence on features and target labels. We compare cPNN and its ablated versions with classical SML models. We also propose a simple way to perform this comparison in a scenario requiring classification to be performed whenever a new data point is generated. In this reproducibility paper, we, therefore, repeat the experiments of our original cPNN paper [9] using the original source code in new contexts (we add new datasets and new models to the benchmark) to generalize our previous findings.[1]

---

[1] The code and the data used in this work are available here for reproducibility: https://github.com/federicogiannini13/cPNN_extended.

The rest of the paper is organized as follows. Firstly, Sect. 2 analyzes the present ideas in literature. Section 3 discusses the motivation and settings of our experiments, while Sect. 4 exhibits the results. Finally, Sect. 5 debates conclusions and future works.

2 Related Works

In this work, we consider three main research areas focused on the challenges of data streams. Figure 1 summarizes the models and their associated areas.

In a data stream classification problem, a data stream is an unbounded sequence of data points $D : d_0, d_1, ..., d_t, d_{t+1}, ...$ with $t \in \mathbb{N}$. Each data point d_t is a tuple $< X_t, y_t >$ where X_t is the feature vector and y_t is the associated label. The assumption is that, after receiving the feature vector X_t, the model must predict the associated label $\hat{y}_t$. The correct label y_t will be available after the prediction and before receiving the new feature vector X_{t+1}. This way, one can apply a *prequential evaluation* mechanism [8] that, whenever a new data point d_t is generated, acts as follows: 1) Infer the label $\hat{y}_t$ by inputting X_t to the model. 2) Update a performance metric using the correct label y_t. 3) Update the model using $< X_t, y_t >$.

In contrast to classical Machine Learning, which assumes that data is independent and identically distributed, a crucial issue is *concept drift* [17]. Data can, in fact, change its distribution over time. A *concept* is the unobservable random process that produces the data points [7]. The concept drift is *virtual* when there is a change in the probabilities $P(X|y)$ or $P(y)$. Notably, this type of change does not affect the class boundary. Conversely, *real concept drifts* affect the probability $P(y|X)$, changing the class boundary. Additionally, concept drifts can be categorized by considering the speed at which they occur. When an *abrupt* concept drift occurs at time t, the concept immediately changes from $t-1$ to t. The new concept gradually or incrementally replaces the previous during *gradual* or *incremental* concept drifts. Concepts can also re-occur over time. Finally, data can exhibit *temporal dependence* when dealing with data streams. In this circumstance, given a data point d_t, $\exists \tau \; P(a_t|b_{t-\tau}) \neq P(a_t)$ where $a_t \in X_t \cup \{y_t\}, b_{t-\tau} \in X_{t-\tau} \cup \{y_{t-\tau}\}$.

SML [3] develops concept drift detectors to detect concept drift automatically [17] and proposes algorithms to adapt to these changes quickly. Tree-based models implement streaming versions of decision trees that incrementally select the node. *Hoeffding Adaptive Tree (HAT)* [2] incorporates the ADaptive WINdowing (ADWIN) change detector [1] for concept drift. Ensemble methods combine predictions using voting strategies. *Adaptive Random Forests (ARF)* [11] use the Poisson distribution for resampling and training on diverse samples. SML usually assumes data points within a concept to be temporal independent. However, *Temporal Augmentation (TA)* [4] has emerged as the first meta-strategy to take temporal dependence into account. Denoting the order by o, TA adds to the feature space of each data point the labels of the previous o data points.

Conversely, CL addresses the forgetting problem when learning from data streams using deep-learning models. It usually assumes the data stream to be

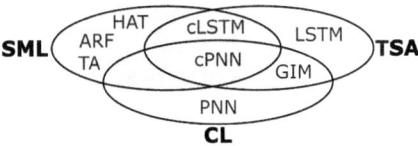

Fig. 1. Summarization of the models considered in this work and their research areas.

split into large batches of data points called experiences. In this scenario, data points do not arrive individually, but a new experience arrives at each timestamp. Each experience introduces a new concept. The goal is to incorporate the knowledge from the new experience without forgetting the previous one. PNNs [23] is a famous solution in this context. During the training on the first experience, a PNN contains a single neural network (called *column*). When a new experience arises, PNN reuses the knowledge gained during previous experiences that could be useful in solving the current problem. To do so, it exploits transfer learning and connects the hidden layers of the different columns. The output of hidden layer i of the column k is computed as in Eq. 1, where W_i and U_i are the weight matrices to be learned. U_i implement transfer learning. When adding a new column, PNN freezes the weights of the previous ones to avoid forgetting. Since, during the inference, it is supposed to know which experience is associated with a specific data point, PNN can select the corresponding column output.

$$h_i^{(k)} = f\left(W_i^{(k)} h_{i-1}^{(k)} + \sum_{j<k} U_i^{(k:j)} h_{i-1}^{(j)}\right) \quad (1)$$

Gated Incremental Memories (GIM) [6] introduces a recurrent version of PNN to manage temporal dependence. It uses LSTM [13], a famous deep-learning model in TSA that works with fixed-size sequences of items. Recent advancements highlighted the forecasting potential of Deep Learning [18]. GIM simplifies PNN by connecting each column only to the preceding one. As expressed by Eq. 2, given a sequence of items and a column k, the new feature vector $X_j^{(k)}$ for the *j-th* item of the sequence concatenates the original feature vector X_j with the LSTM hidden layer's output for the item in column $k - 1$.

$$X_j^{(k)} = X_j \parallel LSTM_j^{(k-1)} \quad (2)$$

GIM works with CL experiences where each sample is a sequence of items, and the different samples within an experience are independent. To address the challenges of continuously learning from mini-batches of data points, managing temporal dependence, handling concept drifts, and preventing forgetting we introduced **Continuous Progressive Neural Networks (cPNN)** [9]. As Fig. 2 shows, it uses cLSTM (a continuous extension of LSTM originally presented by Neto et al. [15]) as GIM's columns. cLSTM manages temporal dependence when learning continuously from a data stream by accumulating data points in fixed-size mini-batches. When the mini-batch is complete, it builds

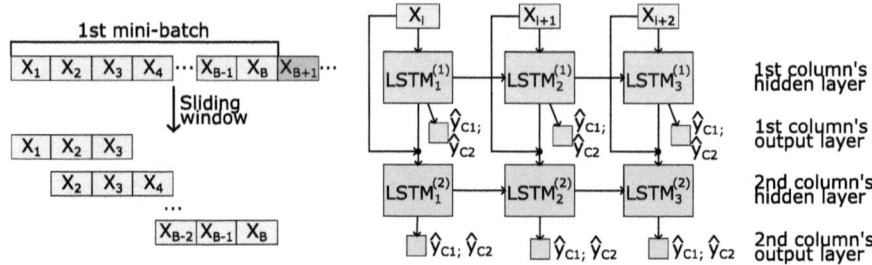

Fig. 2. For each sequence item, cPNN computes scores for each class, averaging scores for the same data point across sequences. Each column concatenates, for each item j, the feature vector X_j and the associated previous column's hidden layer's output.

sequences using a sliding window. Then, it trains a many-to-many LSTM model on the sequences. Given each sequence item, cLSTM outputs a score for each target class. The score $\hat{y}_c(d_t)$ for a given target class c on data point d_t is determined by averaging the scores $\hat{y}_c^s(d_t)$ from all sequences s containing d_t within the mini-batch. For each mini-batch, the loss function is computed as the Binary Cross Entropy, where the prediction for a given target class c on data point d_t is obtained as expressed in Eq. 3.

$$\hat{y}_c(d_t) = Mean\{\hat{y}_c^s(d_t) | d_t \in s\} \qquad (3)$$

cLSTM and cPNN are, thus, periodic classifiers that require mini-batches of data points during both the training and inference phases. This approach can be a reasonable option in various practical situations. As expressed in [22], there are many cases where it is unnecessary to produce predictions immediately. Additionally, online true single-pass learning is rarely preferable and sometimes critical. When the model has seen enough data, the knowledge from new data points can be incorporated into the model with less urgency. Moreover, in a data stream scenario, there are only a few rare cases in which a single training overwhelms the memory of a device. In most cases, storing data points in a buffer is possible by limiting the size of the buffer. Furthermore, utilizing mini-batches can enhance the model's representational capacity when possible. The label for each data point is determined by examining its complete history within the mini-batch, from when the data point is the sequence's final item to when it is the initial one. This allows us to consider both past and future temporal dependence.

Applying GIM prevents forgetting by freezing the weights of the columns associated with the previous concepts. It also exploits transfer learning to boost adaptation after a concept drift by reusing past knowledge. The model combines the knowledge associated with the previous column with the feature coming from the data points. During the training procedure, it learns how to weigh the two. Algorithm 1 outlines the lifecycle of the cPNN [9]. Initially, the architecture consists of a cLSTM (column). The incoming data stream S is buffered into mini-batches of size B (Line 7). Once the mini-batch is complete, a sliding window builds the sequences of size W (Line 10). Next, prequential evaluation [8] is used (Lines 11-13), where the model's predictions are first obtained, and then its performance is assessed on the entire mini-batch. Subsequently, the model is trained on the mini-batch over E epochs. Upon detecting a concept drift, the model processes the accumulated batch (Line 8). After freezing the weights of the last column C, cPNN introduces a new column and connects it to C (Line 16).

Algorithm 1. cPNN training

Input: Data stream S, Batch size B, Epochs E, Window Size W.
1: batch ← []
2: perf ← []
3: model ← new cPNN()
4: drift ← False
5: **for all** (X_t, y_t) in S **do**
6:    **if** drift = False **then**
7:       batch.append((X_t, y_t))
8:    **if** len(batch)=B OR drift=True **then**
9:       **if** len(batch) ≥ W **then**
10:          X,Y ← BuildSequences(batch, W)
11:          pred ← model.predict(X)
12:          perf.append(Evaluate(pred, Y))
13:          model.fit(X, Y, E)
14:       batch ← []
15:    **if** drift=True **then**
16:       model.addColumn()
17:       batch.append((X_t, y_t))
18:    drift ← detectDrift(X_t, y_t)

---

3 Experimental Setting

As highlighted by our cPNN original paper [9], if an abrupt concept drift introduces a classification problem similar to the previous, the Stochastic Gradient Descent (SGD) may be able to adapt to the new concept quickly. Conversely, when the new concept substantially changes the classification problem, SGD may require more steps to converge since it starts from the optimal configuration of the previous concept that could be far from the new one. The goal of cPNN is to boost the adaptation to the new concept by exploiting transfer learning and reusing useful past knowledge. The first objective of this work is to analyze the behaviour of cPNN in the cases of different types of drifts and compare it with SGD on data containing elaborated temporal dependence. Additionally, we are interested in analyzing the behaviour of SML models in the case of temporal dependence. We also introduce the usage of TA to check whether it improves the performance of SML models. Finally, we compare cPNN with SML models.

This Section describes the experimental setting, with Sect. 3.1 explaining the data streams generation, Sect. 3.2 presenting the considered models, and Sect. 3.3 illustrating the hypotheses formulation.

3.1 Data Streams Generation

The most known benchmarks with temporal dependence, as highlighted in [9,24], are unsuitable for our needs. Moreover, existing synthetic data stream generators lack temporal dependence. Hence, we suggest two new generators for synthetic and realistic data. We use the term *boundary function* to indicate a function that determines the boundaries between the classes of a specific classification problem. A *classification function* is, instead, a function that assigns the labels to the points given a boundary function. We indicate as *severe drift* a concept drift where a new concept reverses the labels while maintaining or changing the boundary function. Otherwise, we refer to the drift as *mild drift*. We consider two boundary functions and, thus, four classification functions. Each classification function represents a concept associated with thousands of data points. We combine the four concepts in four different *configurations* by alternating the boundary functions and producing three abrupt concept drifts.

We propose two new generators. *SRWM* generates synthetic data points and produces elaborated temporal dependence on both features and labels. *Weather* approaches a realistic scenario by using data collected from weather sensors.

SRWM Generator. The first generator extends the SineRW generator (*SRW*) presented in [9], which generates points in two dimensions (x_1 and x_2) via a random walk process. It starts with a randomly generated two-dimensional point within the range (0,1). To generate the feature values of a given point, a random walk is performed, where a random value is incrementally added to the previous value [21]. These random walks are drawn from a specific distribution and have a maximum magnitude of 0.05, with a randomly assigned sign. The sign is reversed to prevent feature values from exceeding the (0,1) range if adding the random walk could cause the new value to go beyond this range. In this manner, each feature can be viewed as a time series. To identify the boundaries of the classes, it utilizes the two SINE generator's boundary functions [7] defined in Eq. 4. Each boundary function results in two binary classification functions: one assigns 1 to points meeting the condition obtained by replacing the equality with the inequality ≥ 0, while the others are assigned 0. The second uses < 0. ◰ generates ◰ and ◳, while ◱ generates ◱ and ◲. Examples of severe drifts are from ◰ to ◳ or from ◱ to ◲. Examples of mild drifts are from ◰ to ◱ or from ◲ to ◳.

$$◰ : x_1 - sin(x_2) = 0 \quad ◱ : x_1 - 0.5 - 0.3\, sin(3\,\pi\, x_2) = 0 \tag{4}$$

SRW generator injects temporal dependence in the features of the data points. However, the label of a specific data point can be inferred by looking only at that single data point, without considering sequences. To complicate the classification problem, we incorporate temporal dependence into the data points' labels and produce SineRW Mode (SRWM). The updated label at time t, represented as y'_t, is computed using Eq. 5, where y_t denotes the binary label assigned by SRW. We strengthen the temporal connections between data points by considering also the past labels. This way, a data point's label cannot be inferred only using its features. For each concept, we generate 50k data points.

$$y'_t = MODE(y_t, y_{t-1}, y_{t-2}, y_{t-3}, y_{t-4}) \quad (5)$$

Weather Generator. The second generator starts from the Weather dataset [10] to consider realistic scenarios. Weather is a dataset from the Agricultural Research Service, the U.S. Department of Agriculture's chief scientific in-house research agency. It contains detailed hydrometeorological data from the mountain rain-to-snow transition zone from 2004 through 2014 from the Johnston Draw watershed ($1.8km^2$) in southwestern Idaho. Data includes continuous hourly hydrometeorological variables from two stations across a 372m elevation gradient on the north- and south-facing slopes, resulting in 96432 data points for each station. More precisely, we use the following features: A: Air temperature in °C, RH: Relative Humidity in percentage, w_s: Wind speed in ms^{-1}, w_d: Wind direction from 0° to 360°, T_d: Dew point Temperature in °C. Water vapor pressure is not considered due to its perfect positive correlation with the dew point temperature T_d. We replicate a classification problem by using a specific feature to construct binary labels. We choose this feature after conducting an analysis of the correlations between the features and selecting the most correlated one. In this way, we produce a classification problem that allows for the prediction of the label using the remaining features. We select the first station and choose A as the target feature since it has the following Pearson's R correlation coefficients: 0.7 with RH, 0.6 with T_d, -0.2 with w_s, and -0.3 with w_d. We scale the different features to facilitate the deep-learning learning process.

After choosing the target feature, we create two binary classification functions as expressed in Eqs. 6 and 7 to incorporate elaborate temporal dependence. X_t is the feature vector of the data point at timestamp t, A_t is the value of the target feature. We obtain two further functions (F_{1-}, F_{2-}) by inverting the labels of F_{1+}, and F_{2+}. We can imagine two boundary functions: F_1 compares the current data point with the previous one, F_2 with the median of the previous ten. Examples of severe drifts are from F_{2+} to F_{1-} or from F_{1+} to F_{2-}. Examples of mild drifts are from F_{1+} to F_{2+} or from F_{2-} to F_{1-}. The data set is split into four concepts of the same length, each assigned to a classification function.

$$F_{1+} : y(X_t) = \begin{cases} 1, & \text{if } A_t > A_{t-1} \\ 0, & \text{otherwise} \end{cases} \quad (6)$$

$$F_{2+} : y(X_t) = \begin{cases} 1, & \text{if } A_t > Median(A_{t-10}, ..., A_{t-1}) \\ 0, & \text{otherwise} \end{cases} \qquad (7)$$

3.2 Models and Scenarios

Our experiments compare cPNN and SML models with and without TA. We label SML models with TA as SML_T, and the classical ones without TA as SML_C. We use two SML_C models: *ARF* and *HAT*. Additionally, we create variants with TA: ARF_T and HAT_T. We also test the following *ablated versions* of cPNN to investigate the need for its different components (as also done by [9]):

- *cLSTM*: it directly applies cLSTM without the GIM architecture. After a concept drift, it continues training on the new concept, starting from the optimal parameters of the previous concept. Therefore, it does not avoid forgetting and is unaware of concept drift.
- Multiple cLSTMs (*mcLSTM*): a cPNN in which each column does not consider the previous column's hidden layer output. It avoids, thus, forgetting but does not use transfer learning.

We must consider that SML and cPNN models work differently. cPNN and its ablated versions perform inference and training after accumulating a mini-batch of data points. We call this setting **periodic scenario**. Conversely, we call **anytime scenario** the setting where classification must be performed whenever a new data point is generated. While SML models can perform learning on each data point, cLSTM (and thus mcLSTM and cPNN) learning is designed for mini-batch processing and cannot be altered. Traditional deep-learning models typically run training on fixed-size mini-batches rather than individual data points. This approach facilitates the Stochastic Gradient Descent algorithm to produce more precise gradient estimates and enables parallel processing of data points [12]. We preserve the training phase with mini-batches intact. Nevertheless, we can adjust the inference phase to operate as an anytime classifier. Instead of waiting for the mini-batch to be complete before predicting each data point, cLSTM can instantly predict each data point as generated in the data stream. To achieve this, cLSTM can rely solely on the initial sequence in which the data point appears. Let W be the window size. To classify the data point at time t, the model must buffer the data points from $t - W + 1$ to $t - 1$. This way, cLSTM keeps a **periodic learner** but becomes an **anytime classifier** like SML models and can, thus, make predictions whenever a new data point is generated.

cPNN models hyperparameters are chosen as follows after executing the preliminary experiments (as in our original cPNN paper [9]). Epochs number E: 10, mini-batch size B: 128, learning rate: 0.1, hidden layer size: 50. For Weather, the window size is set to 11 to incorporate the temporal dependence of F_2. SML models use the default parameters, while the order of TA is set to consider the same number of data points of cPNN models. To focus on the models' reaction to drifts, we are supposed to know the exact time of concept drifts. This can result from applying a concept drift detector with 100% accuracy.

3.3 Hypotheses Formulation

We formulate the following research hypotheses:

- **H1**: cPNN adapts to new concepts more quickly than its ablated versions if the drift is severe. Verification: cPNN statistically outperforms its ablated versions after encountering the first part of a concept following a severe drift.
- **H2**: cPNN adapts to new concepts more quickly than its ablated versions if the drift introduces an already-seen boundary function. Verification: cPNN statistically outperforms its ablated versions in the initial part of a new concept that introduces an already-seen boundary function.
- **H3**: cLSTM quickly adapts to a new concept similar to the previous one. Verification: the cPNN performance is not significantly better than cLSTM after encountering a new boundary function following a mild drift.
- **H4**: an SML_C model cannot learn in temporal dependence. Verification: after the initial part of the concepts and upon its end, each SML_C model is statistically outperformed by at least one SML_T and one cPNN model.
- **H5**: SML_T models improve the taming of temporal dependence. Verification: each SML_T model statistically outperforms its version without the TA after the initial part and the end of the concepts.
- **H6**: cPNN models outperform SML_T models, even though the learning phase could be slower. Verification: cPNN or one of its ablated version statistically outperform all other models at the end of each concept.

The comparison between cPNN models is performed in the periodic scenario since it is where cPNN is designed to operate. The training goal is, in fact, not aligned with the inference one in the anytime scenario since the training loss function is computed by averaging the scores of the same data point over a mini-batch. In contrast, inference is computed on single data points. We, thus, verify H1, H2, and H3 in a periodic scenario. H4, H5, and H6 are verified in an anytime scenario since SML models are meant to deal with it.

Since labels can be unbalanced, we use Cohen's kappa score as a performance metric. Cohen's Kappa is a statistical measure used to quantify agreement between two raters [20]. It is usually applied to evaluate the agreement between the true labels and those a learned model assigns. It is calculated as $K = \frac{P_0 \cdot P_e}{1 - P_e}$ where P_0 is the observed agreement, and P_e is the expected agreement by chance. Values range from -1 to 1, with 1 indicating perfect agreement, 0 indicating no agreement beyond chance, and negative values indicating disagreement. The interpretation of Cohen's Kappa values is generally as follows: (0, 0.2) slight agreement, [0.2, 0.4) fair agreement, [0.4, 0.6) moderate agreement, [0.6, 0.8) substantial agreement, [0.8, 1) almost perfect agreement. We compute it using the prequential evaluation [8]. For each model in the anytime scenario, whenever a new data point d_t is generated, we take the prediction, update the score and then update the model. Scores are concept-specific and reset after concept drifts. The score at time t considers all the predictions made by the model from the first data point following the last drift to d_t. In contrast, whenever a mini-batch b_t is filled up in the periodic scenario, we take the predicted labels

Table 1. Average Cohen's Kappa scores of the cPNN ablation study in the periodic scenario. M stands for a concept that introduces mild drift, and S for severe drift.

	SRW (from the 2nd concept onwards)						Weather (from the 2nd concept onwards)					
	◤ (M) start end	◢ (S) start end	◣ (M) start end	F_{2+} (M) start end	F_{1-} (S) start end	F_{2-} (M) start end
cLSTM	.43 .60	.73 .86	.44 .61	**.70** .76	.30 .54	.74 .79
cPNN	.40 .61	**.83 .90**	**.62 .72**	.67 .76	**.69 .73**	**.81 .83**
mcLSTM	.34 .49	.67 .84	.34 .49	.39 .59	.23 .52	.43 .60
	◣ (S) start end	◢ (M) start end	◤ (S) start end	F_{2-} (S) start end	F_{1-} (M) start end	F_{2+} (S) start end
cLSTM	.43 .60	.79 .89	.40 .63	.40 .62	.61 .68	.53 .69
cPNN	.42 .61	**.82 .90**	**.61 .73**	**.65 .75**	**.69 .72**	**.80 .82**
mcLSTM	.33 .51	.67 .84	.33 .50	.41 .61	.25 .53	.43 .60
	▶ (M) start end	◣ (S) start end	◢ (M) start end	F_{1+} (M) start end	F_{2-} (S) start end	F_{1-} (M) start end
cLSTM	.65 .86	.45 .62	.81 .90	**.60 .67**	.49 .65	.60 .68
cPNN	.71 .85	**.58 .69**	**.85 .91**	.57 .64	**.74 .77**	**.66 .67**
mcLSTM	.71 .85	.33 .49	.68 .84	.26 .50	.47 .62	.32 .56
	◢ (S) start end	◣ (M) start end	▶ (S) start end	F_{1-} (S) start end	F_{2-} (M) start end	F_{1+} (S) start end
cLSTM	.66 .86	.44 .63	.69 .85	.23 .48	.73 .7817	.39 .60
cPNN	**.73 .85**	**.59 .69**	**.86 .90**	**.57 .64**	.74 .76	**.65 .67**
mcLSTM	.68 .84	.33 .450	.71 .86	.27 .50	.47 .62	.33 .55

of all the accumulated data points, compute the Cohen's Kappa score on them, and then update the model. We then average the scores from the first mini-batch after the last concept drift to b_t.

As we made in the cPNN paper [9], for each concept, we consider the performance after its first part (*start*) and after its end (*end*). *start* is represented by the first 50 mini-batches and is useful to evaluate how the models react to the drift. In the anytime scenario, we consider the corresponding number of data points ($50 \times B$ where B is the mini-batch size). *end* allows us to evaluate how the models perform on the entire concept. All the experiments are executed ten times. To assess whether one model outperforms another, we conduct a statistical hypothesis test with $\alpha = 0.05$, considering the performance of the ten executions. We first conduct a Shapiro-Wilk test to check for normality. If we cannot reject the null hypothesis for both distributions, we conduct a Welch's t-test. Otherwise, we run a Wilcoxon signed-rank test. We perform a one-sided test in both cases.

4 Results

Table 1 shows the comparison results between cPNN and its ablated versions in the periodic scenario. Since the models share the identical architecture on the first concept, we report the comparison from the second concept onwards.

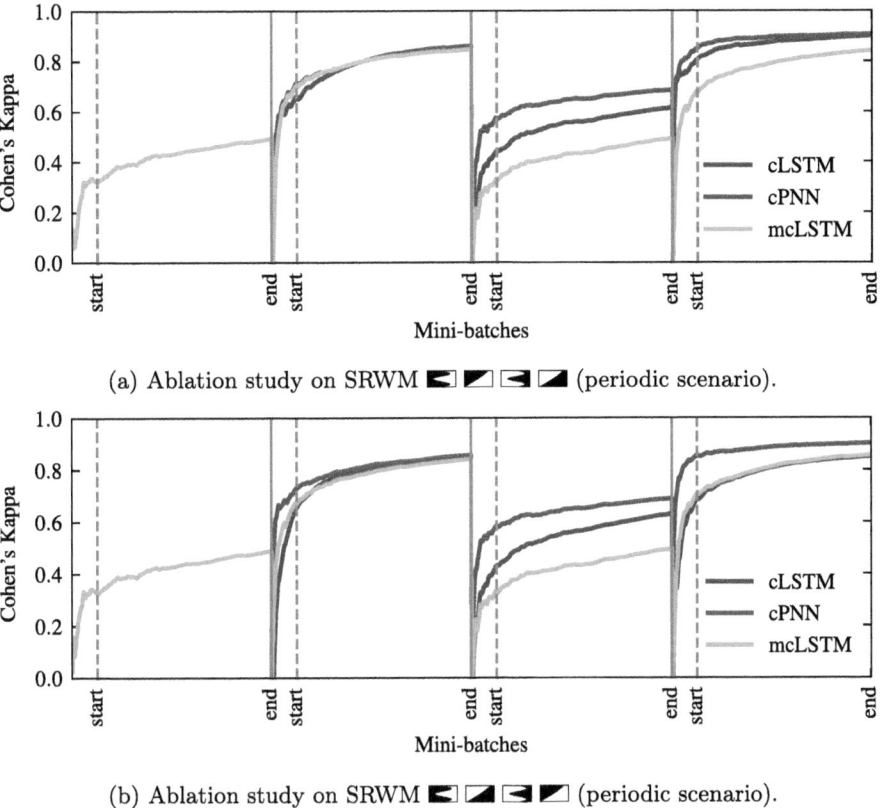

Fig. 3. Ablation study's average Cohen's Kappa scores on two configurations of SRWM. Experiments are executed 10 times. Scores are reset after each drift. cPNN always outperforms the ablated versions in case of severe drift or when the new concept introduces an already-seen boundary function.

For each concept of each configuration, we report the mean of the Cohen's Kappa scores achieved by the models on the 10 iterations after the first 50 mini-batches (start) and the last mini-batch (end). We highlight the statistically best-performing model in bold. Figures 3 and 4 report the complete average Cohen's Kappa evolution on two configurations of SRWM and Weather. The x-axis represent the different mini-batches of the data stream. The values on the y-axis are the Cohen's Kappa scores of the models considering the predictions on the mini-batches from the first following the last drift to the current one. Table 2 presents the results of the comparison between SML_C models, SML_T models, and the maximum performance achieved by cPNN or one of its ablated versions in the anytime scenario. We only present the maximum performance among the cPNN versions because the loss function doesn't align with the inference objective in the anytime scenario. Consequently, the top-performing version of cPNN in the periodic scenario isn't necessarily the optimal choice in the any-

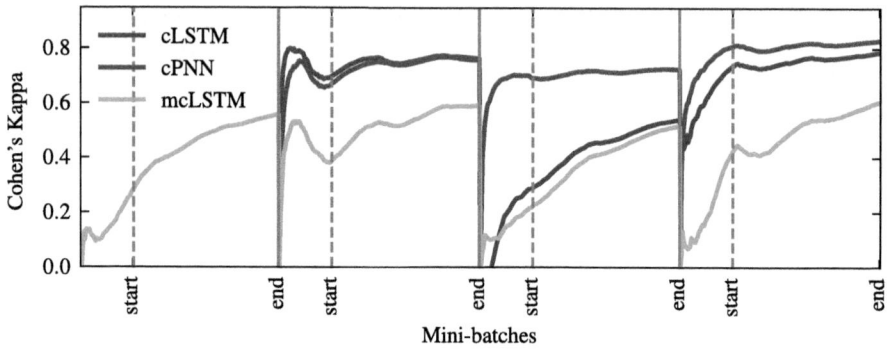

(a) Ablation study on Weather F_{1+} F_{2+} F_{1-} F_{2-} (periodic scenario).

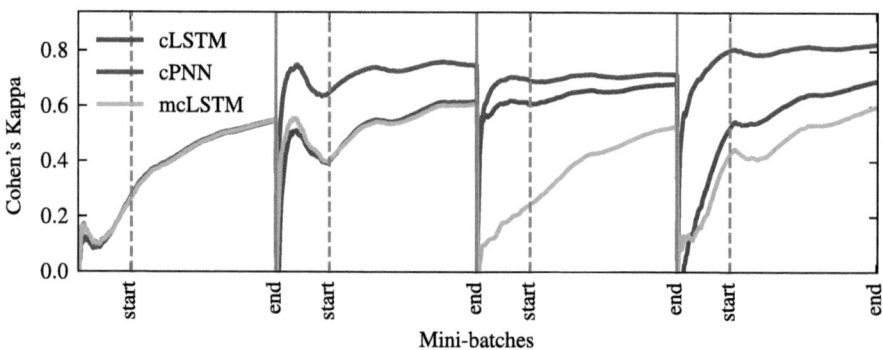

(b) Ablation study on Weather F_{1+} F_{2-} F_{1-} F_{2+} (periodic scenario).

Fig. 4. Ablation study's average Cohen's Kappa scores on two configurations of Weather. cPNN always outperforms the ablated versions in case of severe drift or when the new concept introduces an already-seen boundary function.

time scenario. This could result in applying an ensemble with cLSTM, cPNN, and mcLSTM. We also report the performance at the end of the first concept since it is an object of study. Figures 3 and 4 show the comparison between the maximum performance among the cPNN versions, SML_C, and SML_T. For simplicity, we only report ARF and ARF_T since they outperform HAT and HAT_T, respectively. The x-axis represents the single data points of the data stream. On the y-axis, we report the Cohen's Kappa scores of the models considering the data points from the first following the last drift to the current one. Finally, Table 3 reports the summarization of our hypothesis tests for the four concepts of each configuration.

Concerning the ablation study, hypotheses **H1** and **H2** are quite always verified. This finding proves the effectiveness of cPNN in exploiting transfer learning to better adapt to drifts when the drift is severe or when the boundary function is known. The advantages of using cPNN in the case of severe drift are more evident in Weather configurations. This is because SRWM's classification

Table 2. Average Cohen's Kappa scores of the different models in the anytime scenario.

	SRW							Weather							
	end	start	end	start	end	start	end	end	start	end	start	end	start	end	start	end
	▶		⊏		▲		◀	F_{1+}	F_{2+}		F_{1-}		F_{2-}		
ARF	.36	.26	.33	.23	.36	.23	.33	.17	.48	.49	.19	.19	.48	.51		
ARF_T	.73	.74	.74	.73	.74	.74	.74	.30	.74	.75	.34	.33	.76	.76		
HAT	.35	.26	.34	.17	.34	.24	.33	.16	.38	.40	.15	.17	.42	.42		
HAT_T	.60	.63	.66	.63	.65	.60	.64	.29	.59	.59	.29	.30	.61	.60		
cPNN	.78	.33	.50	.69	.80	.47	.51	.50	.73	.77	.57	.57	.75	.79		
	▶		◀		▲		⊏	F_{1+}	F_{2-}		F_{1-}		F_{2+}		
ARF	.36	.21	.33	.23	.35	.20	.33	.18	.27	.43	.20	.20	.36	.47		
ARF_T	.73	.74	.74	.73	.74	.74	.74	.30	.73	.75	.35	.33	.75	.76		
HAT	.35	.21	.32	.24	.35	.19	.33	.15	.28	.38	.18	.18	.37	.41		
HAT_T	.60	.57	.66	.63	.67	.64	.66	.29	.60	.58	.29	.30	.60	.58		
cPNN	.78	.33	.50	.64	.79	.43	.51	.49	.62	.71	.57	.62	.74	.76		
	⊏		▶		◀		▲	F_{2+}	F_{1+}		F_{2-}		F_{1-}		
ARF	.35	.23	.34	.23	.33	.23	.35	.47	.20	.21	.38	.46	.23	.21		
ARF_T	.74	.74	.74	.74	.74	.73	.74	.74	.30	.34	.75	.75	.38	.37		
HAT	.33	.26	.35	.19	.32	.24	.35	.38	.16	.17	.38	.39	.18	.17		
HAT_T	.59	.65	.62	.53	.61	.63	.66	.58	.27	.31	.53	.57	.34	.32		
cPNN	.38	.58	.81	.47	.53	.69	.84	.54	.51	.60	.70	.71	.54	.63		
	⊏		▲		◀		▶	F_{2+}	F_{1-}		F_{2-}		F_{1+}		
ARF	.35	.17	.35	.23	.33	.21	.35	.47	.15	.19	.50	.49	.22	.21		
ARF_T	.74	.73	.74	.74	.74	.74	.74	.74	.30	.34	.76	.75	.38	.37		
HAT	.33	.18	.34	.23	.33	.19	.34	.38	.12	.16	.42	.40	.15	.16		
HAT_T	.55	.58	.64	.63	.66	.62	.61	.58	.28	.31	.52	.57	.33	.32		
cPNN	.38	.57	.80	.45	.52	.54	.79	.54	.45	.47	.73	.76	.40	.50		

functions are easier to learn, and cLSTM and mcLSTM can react more quickly to the drift. Conversely, Weather contains more complex data from a realistic scenario. When the drift is mild, and the boundary function is new, cLSTM can start the concept with a higher performance than cPNN. These are the cases in which cLSTM capitalizes its previous optimal settings to optimize the new loss function. This finding allows us to confirm the hypothesis **H3**. Secondly, if the knowledge of the old concept is still valid, cPNN may slow down when calibrating its weights. However, at the end of the concept, cPNN outperforms cLSTM or obtains the same performance. Additionally, mcLSTM is usually the less-performing cPNN version. Using past knowledge is crucial to adapt to the new concept. cPNN does it with transfer learning. cLSTM does it by starting from the optimal setting of the previous concept. This could be enough only when the new concept is similar to the previous one.

When comparing cPNN and SML models, SML_C models cannot learn, reach poor performance, and struggle. Hypotheses **H4** and **H5** are, in fact, always

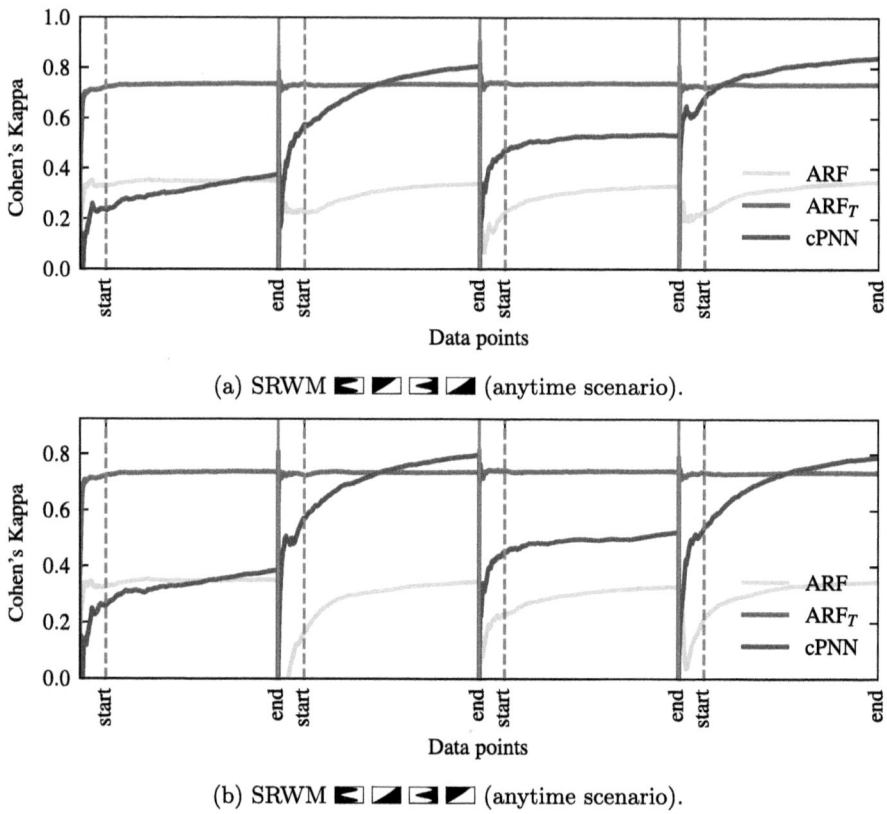

Fig. 5. Average Cohen's Kappa scores on two configurations of SRWM in the anytime scenario. cPNN is computed as the maximum performance between cLSTM, cPNN, and mcLSTM. ARF cannot learn efficiently. The best-performing cPNN version can outperform ARF_T on the concepts associated with ◩.

verified, and SML_C models are always outperformed by at least its version with TA and one cPNN model. Temporal augmentation is crucial to improve the SML models' performance. **H6** is verified in most cases. On SRWM, the best-performing cPNN version can outperform the best-performing SML_T model on the concepts associated with ◩. On the most complex function (◤), the best-performing cPNN version cannot reach the performance of SML_T but constantly outperforms SML_C models without TA. When we move to a more realistic approach like Weather, the best-performing cPNN version outperforms SML_T models on F_1.

One key finding is that SML_T models perform poorly on the only boundary function that reduces the number of data points with label changes compared to the previous. F_1, in fact, retains the previous label for only 68% of data points, resulting in SML_T models achieving Cohen's Kappa scores lower than 0.4 for related concepts. Conversely, F_2 and the SRWM functions keep the previous

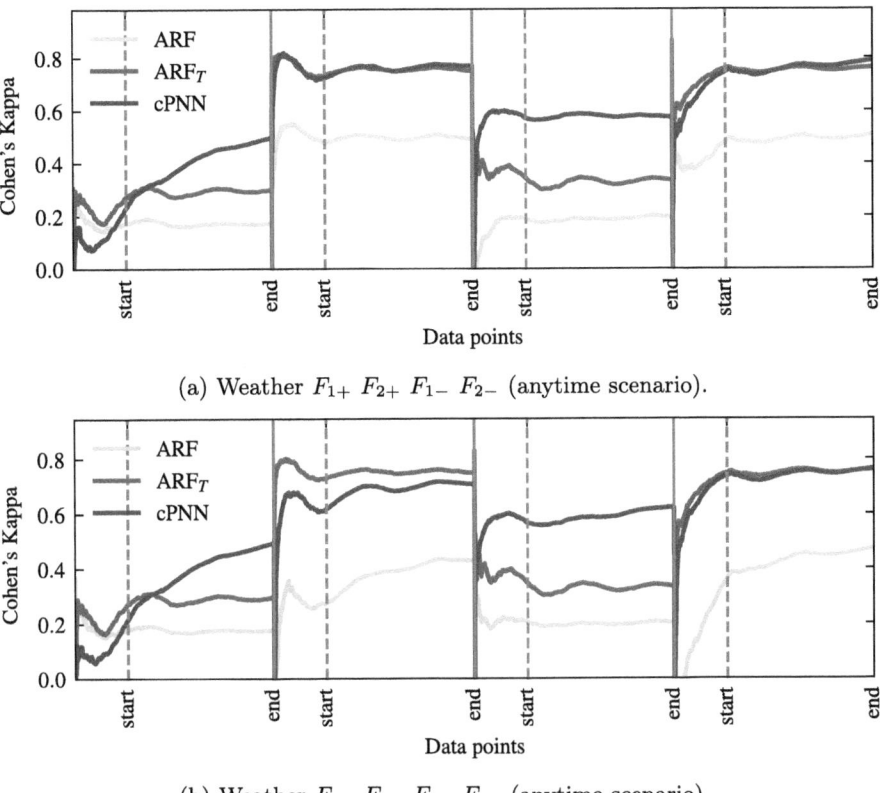

(a) Weather F_{1+} F_{2+} F_{1-} F_{2-} (anytime scenario).

(b) Weather F_{1+} F_{2-} F_{1-} F_{2+} (anytime scenario).

Fig. 6. Average Cohen's Kappa scores on two configurations of Weather in the anytime scenario. cPNN is computed as the maximum performance between cLSTM, cPNN, and mcLSTM. ARF cannot learn efficiently. The best-performing cPNN version outperforms ARF_T on F_1.

label for nearly 90% of data points. ARF_T achieves scores exceeding 0.72 on these functions. This suggests that TA may be biased on the previous label. This suggestion is confirmed by Fig. 5, where ARF_T performance on SRWM keeps unaltered over the entire data stream. On Weather (Fig. 6), instead, it takes longer to converge to the final score, and the drift has a clear impact. Additionally, when the label y_t of data point d_t is unavailable before receiving X_{t+1}, TA cannot be applied since it relies on the previous labels.

Table 3. Hypotheses tests results for SRWM and Weather. We indicate whether the hypothesis is verified (✓) or not (×) for each of the four concepts, across every generator, configuration, and hypothesis. When a hypothesis does not apply to a specific concept we use "-".

		H1	H2	H3	H4	H5	H6
SRWM	▶◣◢◥	- - ✓ -	- - ✓✓	- ✓ - -	✓✓✓✓	✓✓✓✓	✓× ✓×
	▶◥◢◣	- × - ✓	- - ✓✓	- - - -	✓✓✓✓	✓✓✓✓	✓× ✓×
	◣▶◥◢	- - ✓ -	- - ✓✓	- ✓ - -	✓✓✓✓	✓✓✓✓	× ✓× ✓
	◣◢◥▶	- ✓ - ✓	- - ✓✓	- - - -	✓✓✓✓	✓✓✓✓	× ✓× ✓
Weather	$F_{1+}\ F_{2+}\ F_{1-}\ F_{2-}$	- - ✓ -	- - ✓✓	- ✓ - -	✓✓✓✓	✓✓✓✓	✓✓✓✓
	$F_{1+}\ F_{2-}\ F_{1-}\ F_{2+}$	- ✓ - ✓	- - ✓✓	- - - -	✓✓✓✓	✓✓✓✓	✓× ✓×
	$F_{2+}\ F_{1+}\ F_{2-}\ F_{1-}$	- - ✓ -	- - ✓✓	- ✓ - -	✓✓✓✓	✓✓✓✓	× ✓× ✓
	$F_{2+}\ F_{1-}\ F_{2-}\ F_{1+}$	- ✓ - ✓	- - × ✓	- - - -	✓✓✓✓	✓✓✓✓	× ✓✓✓
	Verified cases	11	15	4	32	32	19
	Total cases	12	16	4	32	32	32

5 Conclusion

This study extensively analyzed various streaming models' behaviour in handling concept drifts with complex temporal dependence. The examined models included classical SML, SML with TA, and cPNN. We initially used a synthetic data generator to simulate temporal dependence, then transitioned to real data for a more realistic scenario. An ablation study compared cPNN's response to drift with SGD, showing cPNN's adeptness, particularly in severe drift or previously encountered functions. Comparisons with classical SML models highlighted cPNN's efficiency in managing temporal dependence, outperforming SML and SML with TA models. Additionally, Classical SML models were not able to learn efficiently.

Two main limitations of cPNN are its complexity grows linearly with the number of concepts and the difficulty of hyperparameter selection in data streams. Since assuming the exact timestamps of drifts is unrealistic, future research could explore using drift detection methods. Additionally, cPNN is a periodic learner requiring mini-batches. We proposed a first approach to transform it into an anytime classifier. Modifying the loss function could enhance classification at any time. Since our results suggest TA may be biased toward the previous label, a future study can investigate TA's behaviour with complex functions that frequently change labels. Additionally, in the case of delayed labels TA cannot be applied and other methods should be found to manage temporal dependence. Finally, while cPNN is naturally robust to forgetting, we leave the study of forgetting in cLSTM and SML models for future research.

References

1. Bifet, A., Gavaldà, R.: Learning from time-changing data with adaptive windowing. In: SDM, pp. 443–448. SIAM (2007)
2. Bifet, A., Gavaldà, R.: Adaptive learning from evolving data streams. In: Adams, N.M., Robardet, C., Siebes, A., Boulicaut, J.-F. (eds.) Advances in Intelligent Data Analysis VIII, pp. 249–260. Springer Berlin Heidelberg, Berlin, Heidelberg (2009). https://doi.org/10.1007/978-3-642-03915-7_22
3. Bifet, A., Gavaldà, R., Holmes, G., Pfahringer, B.: Machine learning for data streams: with practical examples in MOA. MIT press (2018)
4. Bifet, A., Read, J., Žliobaitė, I., Pfahringer, B., Holmes, G.: Pitfalls in benchmarking data stream classification and how to avoid them. In: Blockeel, H., Kersting, K., Nijssen, S., Železný, F. (eds.) ECML PKDD 2013. LNCS (LNAI), vol. 8188, pp. 465–479. Springer, Heidelberg (2013). https://doi.org/10.1007/978-3-642-40988-2_30
5. Box, G.E., Jenkins, G.M., Reinsel, G.C., Ljung, G.M.: Time series analysis: forecasting and control. John Wiley & Sons (2015)
6. Cossu, A., Carta, A., Bacciu, D.: Continual learning with gated incremental memories for sequential data processing. In: IJCNN, pp. 1–8. IEEE (2020)
7. Gama, J., Medas, P., Castillo, G., Rodrigues, P.: Learning with drift detection. In: Bazzan, A.L.C., Labidi, S. (eds.) SBIA 2004. LNCS (LNAI), vol. 3171, pp. 286–295. Springer, Heidelberg (2004). https://doi.org/10.1007/978-3-540-28645-5_29
8. Gama, J., Sebastião, R., Rodrigues, P.P.: Issues in evaluation of stream learning algorithms. In: KDD, pp. 329–338. ACM (2009)
9. Giannini, F., Ziffer, G., Della Valle, E.: cPNN: continuous progressive neural networks for evolving streaming time series. In: Kashima, H., Ide, T., Peng, W.-C. (eds.) Advances in Knowledge Discovery and Data Mining: 27th Pacific-Asia Conference on Knowledge Discovery and Data Mining, PAKDD 2023, Osaka, Japan, May 25–28, 2023, Proceedings, Part IV, pp. 328–340. Springer Nature Switzerland, Cham (2023). https://doi.org/10.1007/978-3-031-33383-5_26
10. Godsey, S.E., et al.: Eleven years of mountain weather, snow, soil moisture and streamflow data from the rain-snow transition zone-the Johnston Draw catchment, Reynolds Creek Experimental Watershed and Critical Zone Observatory, USA. Earth Syst. Sci. Data **10**(3), 1207–1216 (2018)
11. Gomes, H.M., et al.: Adaptive random forests for evolving data stream classification. Mach. Learn. **106**(9-10), 1469–1495 (2017)
12. Goodfellow, I.J., Bengio, Y., Courville, A.C.: Deep Learning. MIT Press, Adaptive computation and machine learning (2016)
13. Hochreiter, S., Schmidhuber, J.: Long short-term memory. Neural Comput. **9**(8), 1735–1780 (1997)
14. Lange, M.D., et al.: A continual learning survey: defying forgetting in classification tasks. IEEE Trans. Pattern Anal. Mach. Intell. **44**(7), 3366–3385 (2022)
15. Lemos Neto, Á.C., Coelho, R.A., Castro, C.L.d.: An incremental learning approach using long short-term memory neural networks. J. Contr. Autom. Electr. Syst., pp. 1–9 (2022)
16. Lesort, T., Lomonaco, V., Stoian, A., Maltoni, D., Filliat, D., Rodríguez, N.D.: Continual learning for robotics: Definition, framework, learning strategies, opportunities and challenges. Inf. Fusion **58**, 52–68 (2020)
17. Lu, J., Liu, A., Dong, F., Gu, F., Gama, J., Zhang, G.: Learning under concept drift: a review. IEEE Trans. Knowl. Data Eng. **31**(12), 2346–2363 (2019)

18. Makridakis, S., Spiliotis, E., Assimakopoulos, V.: M5 accuracy competition: Results, findings, and conclusions. IJOF **38**(4), 1346–1364 (2022)
19. McCloskey, M., Cohen, N.J.: Catastrophic interference in connectionist networks: The sequential learning problem. In: Psychology of learning and motivation, vol. 24, pp. 109–165. Elsevier (1989)
20. McHugh, M.L.: Interrater reliability: the kappa statistic. Biochemia medica **22**(3), 276–282 (2012)
21. Pearson, K.: The problem of the random walk. Nature **72**(1865), 294–294 (1905)
22. Read, J., Zliobaite, I.: Learning from Data Streams: An Overview and Update. CoRR **abs/2212.14720** (2022)
23. Rusu, A.A., et al.: Progressive Neural Networks. CoRR **abs/1606.04671** (2016)
24. de Souza, V.M.A., dos Reis, D.M., Maletzke, A.G., Batista, G.E.A.P.A.: Challenges in benchmarking stream learning algorithms with real-world data. Data Min. Knowl. Discov. **34**(6), 1805–1858 (2020)
25. Ziffer, G., Bernardo, A., Della Valle, E., Cerqueira, V., Bifet, A.: Towards time-evolving analytics: online learning for time-dependent evolving data streams. Data Sci. **6**(1–2), 1–16 (2023)

Author Index

B
Baralis, Elena   101
Basterrech, Sebastián   128
Boystov, Andrey   57

C
Carbajo, Ricardo Simón   3
Chandra, Joydeep   20
Chen, Xue   37
Cordy, Maxime   57

D
de Alfaro, Luca   101
Della Valle, Emanuele   86, 146

F
Fastowski, Alina   74

G
Ghamizi, Salah   57
Giannini, Federico   146
Giobergia, Flavio   101
Goujon, Anne   57
Guo, Jiaqi   37

K
Kasneci, Gjergji   74
Komorniczak, Joanna   111
Ksieniewicz, Paweł   111

L
Lefebvre, Clément   57
Li, Tianpeng   37

O
Ordóñez, Sebastián A. Cajas   3

P
Pan, Lin   37
Pastor, Eliana   101

S
Samanta, Jaydeep   3
Saxena, Shruti   20
Shao, Minglai   37
Simonetto, Thibault   57
Suárez-Cetrulo, Andrés L.   3
Sun, Yueheng   37

T
Traon, Yves Le   57

W
Wang, Wenjun   37

Z
Ziffer, Giacomo   86

GPSR Compliance

The European Union's (EU) General Product Safety Regulation (GPSR) is a set of rules that requires consumer products to be safe and our obligations to ensure this.

If you have any concerns about our products, you can contact us on ProductSafety@springernature.com

In case Publisher is established outside the EU, the EU authorized representative is:

Springer Nature Customer Service Center GmbH
Europaplatz 3
69115 Heidelberg, Germany

Batch number: 08150511

Printed by Printforce, the Netherlands